Boswellin®

The Anti-inflammatory Phytonutrient

Muhammed Majeed, Ph.D.
Vladimir Badmaev, M.D., Ph.D.
S. Gopinathan, Ph.D.
R. Rajendran, M.S.
Todd Norton

FIRST PRINTING

NUTRISCIENCE PUBLISHERS, INC.
Piscataway, New Jersey

The information contained in this book, Boswellin® the anti-inflammatory phytonutrient, is intended for educational purposes only. It is not designed, in whole or in part, as advice for self treatment. Neither is this book, nor any statements by its authors or publisher, to be misconstrued as an endorsement of self treatment.

NUTRISCIENCE PUBLISHERS, INC.

121 Ethel Road West, Unit 6
Piscataway, NJ 08854 USA
Tel: (908) 777-1111
Fax: (908) 777-1443

ISBN: 0-9647856-1-7

ABOUT THE AUTHORS:

Muhammed Majeed, Ph.D., editor, holds a doctorate (1986) in industrial pharmacy from St. John's University in New York. He has over 15 years of pharmaceutical research experience in the United States with leading companies such as Pfizer, Inc., Carter-Wallace, and Paco Research. Dr. Majeed has a broad knowledge of the pharmacological properties of herbal medicines used in Ayurveda, the traditional system of botanical medicine of India. He pioneered the introduction of several herbal remedies for the American and European markets.

Vladimir Badmaev, M.D.,Ph.D., trained in clinical and anatomical pathology at Kings County Hospital and Downstate Medical Center, New York. His Ph.D. degree is in the field of immunopharmacology. He is the author of many articles and a book on traditional medicine, with an emphasis on Ayurvedic and Tibetan medicine.

S. Gopinathan, Ph.D., has a doctorate in botany with over 15 years research experience working on the isolation and identification of plant-derived active compounds. He is a research scientist at Sami Chemicals & Extracts (P) Ltd., Bangalore, India.

R. Rajendran, M.S., has a master's degree in biochemistry and more than 20 years experience in the field of pharmaceuticals, nutriceuticals, and organic synthesis. He is the president of Sami Chemicals & Extracts, (P) Ltd., Bangalore, India.

Todd Norton, has a degree in business administration and over 17 years of diversified experience in the natural products industry.

The authors wish to acknowledge Tang-Sheng Peng, Ph.D., William F. Popin, M.S., and Douglas F. Johnston, B.S. from Nature's Herbs Analytical Chemistry Group for their contribution to the analytical profile section of this book.

FOREWORD

A recent letter appearing in Dr. Paul Donohue's nationally syndicated newspaper column "Good Health" reads,

"Dear Dr. Donohue: I have had rheumatoid arthritis for more than 30 years and recently developed a stomach ulcer. I am no longer able to take aspirin or NSAIDs [nonsteroidal anti-inflammatory drugs including Ibuprofen, Advil, Motrin, Clinoril, Naprosyn, Indocin, Nuprin, Nalfon, Feldene, Ansaid, Bufferin, Rufin and Tolectin]. I am told to take Tylenol, but this is not satisfactory. Any suggestions? -- M.T.

Dr. Donohue's reply,

"Dear M.T: You are truly between a rock and a hard place. Aspirin and the nonsteroidal anti-inflammatories such as ibuprofen and indomethacin make up a most popular group of arthritis-control drugs. Unfortunately, your stomach ulcer makes such drugs off-limits.

And what kind of substitute drugs does Dr. Donohue suggest?

"There are other medicines to try, even with the ulcer, such as Arthropan [an aspirin-containing substitute] or Trilisate [another aspirin derivative]. Or, you can try enteric-coated aspirin.....And your doctor has further choices, including gold, penicillamine or methotrexate [so-called second-line drugs recommended when NSAIDs no longer work well, usually advocated in the hope of preventing all-too-frequent irreversible joint damage & deformity]."

What can M.T. and other ulcer-prone chronic pain sufferers expect from these "substitute" drugs?

"No one can guarantee that substitute drugs won't cause an ulcer flare. As is sometimes the case, you might have to weigh the side effect risk against *the pain of nontreatment*".

"The pain of nontreatment" suggested by Dr. Donohue and unfortunately by many of his physician colleagues is a non-option. It's extremely revealing that once NSAIDs and second line drugs have been tried and failed, conventional allopathic medicine has no legitimate, science-based alternative therapies to recommend. Their option appears to be, "Suffer with chronic debilitating pain or take conventional NSAIDs."

It is estimated there are 40 million arthritic pain sufferers in the United States, most of whom are on aspirin or aspirin substitutes. NSAIDs as a group are one of the most commonly prescribed collection of drugs worldwide. An estimated 10 million victims of arthritis are currently taking *large* doses of NSAIDs on a daily basis (many of whom are on multiple drugs - 4 to 12 drugs simultaneously is not uncommon, often prescribed to counteract familiar side-effects of conventional NSAIDs).

Boswellin® - The Anti-inflammatory Phytonutrient

BOSWELLIA SERRATA EXTRACT:
A SAFE & EFFECTIVE ALTERNATIVE TO CONVENTIONAL NSAIDs

Over the last decade or two a number of healing herbs have been researched, mostly in the medical and scientific communities of Europe and Asia, demonstrating an unexpected ability to improve the symptoms of different arthritics on a level equal to NSAIDs (nonsteroidal anti-inflammatory drugs) - but with no toxic side effects! One such standardized herbal extract nicely and thoroughly presented in this booklet is *Boswellia serrata.*

The major use of *Boswellia serrata* in contemporary medicine is as an anti-arthritic and anti-inflammatory plant-derived extract. Its anti-inflammatory actions, supported by solid science, are indicated in clinical medicine for the treatment of rheumatoid arthritis, osteoarthritis, juvenile rheumatoid arthritis, soft tissue rheumatism, gout, low back pain, myositis and fibrositis, and are attributed primarily to the presence of boswellic acids (These acids have been carefully identified and *standardized* in the Boswellin® product, a production step necessary in order to assure the health consumer that, in fact, the active ingredients are in each tablet, capsule or powder). The therapeutic action of a standardized *Boswellia serrata* includes: reduction of joint swelling, restoration and improvement of blood supply to inflamed joints, pain relief, increased mobility, amelioration of morning stiffness, steroid-sparing effect and general improvement in the quality of life.

Boswellia serrata extract, while demonstrably effective in the treatment of different chronic inflammatory conditions, *produces none of the common side-effects associated with conventional NSAIDs and second-line drugs!* There are no reports of GI ulceration, no GI bleeding, no GI perforation and no kidney disease. The absence of NSAID-related toxicity is the result of its unique site of action: namely, the blocking of the enzyme lipoxygenase and the resulting decrease of pro-inflammatory leukotrienes. On the other hand *Boswellia serrata* extract does not affect cyclo-oxygenase and the synthesis of prostaglandins.

Many safe and effective natural alternatives are now within our reach. As a consequence, no one in chronic pain need any longer face the singular choice of long-term NSAID medications as the only pain remedy. There are times, or course, when potent drugs may be necessary. But my experience as a physician and investigator compels me to recommend that potent pain medications, especially for people over the age of 60, should be used as a last resort, and as sparingly as possible. By educating yourself, either as a health professional or a victim of chronic pain, about the remarkable world of proven natural alternatives, including *Boswellia serrata* in this fine booklet, millions of people will to discover the benefits afforded them through safe, time-tested, natural remedies."

James Braly, M.D.
Boca Raton, Florida
February, 1996

INTRODUCTION

INTRODUCTION

1. History and traditional uses

Boswellia serrata (N.O. Burseraceae) is a large, branching, d e c i d u o u s tree which grows abundantly in the dry, hilly parts of India. It is known as "Dhup," Indian Frankincense or Indian Olibanum.[2,4] The gum resin exudate of *Boswellia serrata,* known in the vernacular as "Salai guggal," has been used in the Ayurvedic system of medicine for the treatment of rheumatism, respiratory diseases, and liver disorders.[1,2,3] The constituents of the resin include: essential oil, terpenoids, and gum. The essential oil of *Boswellia serrata* is used in many oriental perfumes and is closely related to the incense of Biblical renown.[13] The essential oil is also an effective insecticide.[26]

The two important treatises on Ayurveda, authored by the ancient writers, Sushruta Samhita[41] and Charak Samhita[42] describe the antirheumatic activity of guggals (gum resins of certain trees), especially those of *Boswellia serrata.* The guggal from *Commiphora mukul,* (Gugulipid®) another member of the same order, which has been commonly used in Ayurvedic medicine, was shown to have excellent antihyperlipidemic properties[9] and valued in the treatment of obesity. In addition to rheumatoid arthritis, the gummy exudate of *Boswellia serrata* is recommended in Ayurvedic texts as a remedy for diarrhea, dysentery, ringworm, boils and other afflictions.[4,43] Boswellin®, the product trade mark of Sabinsa Corporation, is the selectively fractionated principle obtained from the gummy exudate of *Boswellia serrata.*

The oleogum resin is tapped from the tree by scraping away a portion of the stem bark, usually 15 to 20 cm wide. The practice currently employed is to make transverse incisions in the upper and lower portions of the tree trunk and the bark in between is peeled off. The gum exudate is collected for the next 10-12 days. The average yield of resin per tree is about 1 kg a year.[14]

In Ayurvedic medicine, the gum is described as being sweet, bitter, hot; antipyretic, lowering blood glucose levels and antidysenteric.[3] It is believed to be useful in the treatment of skin and blood diseases, fevers, cardiovascular disorders, neurologic disorders, mouth sores, vaginal discharges, rheumatism, dysentery, diabetes and diseases of the testes.[3] The gum has been used in Ayurvedic medical practice as an astringent, hepatic stimulant (supporting liver functions), in improving blood circulation and in the preparation of an ointment for sores.[3] It is also prescribed with clarified butter in syphillis and in the treatment of jaundice and hemorrhoids.[3] It is often used as a stimulant in pulmonary diseases, cough with copious amounts of sputum and chronic laryngitis and also used externally in the form of a fumigant. If massaged into the scalp in the form of an oily solution, it is believed to enhance the growth of hair.[3] It has also been used to improve appetite and alleviate general weakness and debility.[35] It is also prescribed in menorrhea, dysmenorrhea, gonorrhea and liver diseases.[4,43] The oil extracted from the gum resin is prescribed in the treatment of gonorrhea. A paste made of the gum resin with coconut oil or lemon juice is applied to ulcers, indolent swellings, carbuncles, boils and ringworm. In most conditions, the doses prescribed are:[44]

Gum resin: 2-3 gm
Oil: 1-1.5 ml
Bark decoction: 56-112 ml

In the Ayurvedic texts of Charaka and Sushruta, the oleogum resin is recommended in combination with other drugs for the treatment of snake-bite and scorpion-sting. In the Unani system of medicine it is considered an astringent to the bowels and valued for its expectorant properties, and may induce vomitting.[3] It is useful in this system of medicine in the treatment of intestinal inflammatory conditions, bronchitis, asthma, cough and sore throat. It also heals wounds, strengthens the teeth, is invigorating and used externally in the treatment of boils and scabies. It is applied externally to treat ophthalmic inflammatory conditions.[3]

2. Pharmacological effects and contemporary uses in medicine

The major use of *Boswellia serrata* in contemporary medicine is as an anti-arthritic and anti-inflammatory pharmacologic agent.[5,6,7,21,22] Boswellin®, a standardized alcoholic extract containing pentacyclic triterpene acids, a complex of boswellic acids, is used in these applications. Boswellic acids have also been shown to possess antihyperlipidemic and anti-atherosclerotic properties.[27,28,45] A non-acidic, oil fraction obtained from the gum resin is reported to possess analgesic and psychopharmacological properties.[8] Crude ethanolic extracts of *Boswellia serrata* were also found to possess hepatoprotective (protecting the liver) properties against hepatitis.[20]

Anti-arthritic properties

The gum resin collected from the *Boswellia serrata* tree was shown to be very effective in the treatment of inflammatory and arthritic conditions.[5,6,7,21,22] The anti-inflammatory and anti-arthritic properties of the crude material are attributed to the presence of β-boswellic acid and other related pentacyclic triterpene acids.[15] The active principles are obtained from the gum resin by selective extraction, fractionation and concentration under vacuum.[16]

A fraction of Salai guggal, (trade name Sallaki®), prepared from the gum resin exudate of *Boswellia serrata* was the second therapeutic agent of Ayurvedic origin, to be marketed as a standardized pharmaceutical product. *Rauwolfia serpentina* is considered the source of the first standardized drug of Ayurvedic origin, Reserpine, a result of Col. R.N. Chopra's work in the 1940's and 1950's.

The development of Sallaki® was based on extensive research carried out at the Regional Research Laboratory (RRL, Council of Scientific and Industrial Research), Jammu, India.[10] As a result of this research Sallaki® has been indicated in: rheumatoid arthritis, osteoarthritis, juvenile rheumatoid arthritis, soft tissue rheumatism, low back pain, gout, cervical spondylosis, myositis and fibrositis.[10,11] The therapeutic action of Sallaki® includes: reduction in joint swelling, increased mobility, steroid sparing action (less steroids required in combined treatment), amelioration of morning stiffness and general improvement in quality of life.[10,11] These therapeutic effects were observed primarily with patients suffering from osteoarthritis and rheumatoid arthritis.[10,11]

Rheumatoid arthritis is a relatively common disease, a great crippler, causing immense physical suffering. It is not usually possible to cure this disease. However, it is possible to alleviate physical pain, increase mobility and prevent further tissue injury through proper treatment.[19] The widely used conventional drugs for the treatment of this condition produce undesirable side effects, including gastric irritation leading to the inflammatory condition of gastrointestinal mucosa.

Anti-inflammatory properties

The anti-inflammatory properties of boswellic acids are among the most sought after pharmacological properties. The 100-year-old drug, aspirin, a well known anti-inflammatory agent, which belongs to the group of non-steroidal anti-inflammatory drugs, or NSAID's, is known to produce some side effects, such as gastric irritation. The oleogum resin from *Boswellia serrata* was investigated by scientists at the RRL, for efficacy against inflammatory disorders, with the hope of identifying herbal based anti-inflammatory products that produced no adverse and undesirable side effects.[12] It was found that Salai guggal, while being an effective anti-inflammatory compound, produced no untoward side effects common to most anti-inflammatory drugs, such as kidney failure, peptic ulcer or gastrointestinal hemorrhage. Salai guggal has also been recognized as safe for use during pregnancy.[10]

An inflammatory reaction is usually a response to tissue injury, which is characterized by five symptoms: redness, heat, pain, swelling and decreased function of the affected part of the body. These symptoms are caused by a complex body response involving a broad range of pharmacologically active substances, e.g. bradykinins, histamines, prostaglandins, thromboxanes, hydroxy-fatty acids, leukotrienes, lysosomal enzymes and lymphokines. Boswellic acids, the active anti-inflammatory principles of *Boswellia serrata*, appear to be specific inhibitors of leukotriene formation, specifically inhibiting the activity of the enzyme which leads to their formation. This forms the biochemical basis of their anti-inflammatory action. Boswellic acids are therefore effective in the prevention and/or control of inflammatory processes, which are typically characterized by increased leukotriene formation.[75] They consequently find extensive application in human and veterinary medicine. A recent report advocates the use of an extract of *Boswellia serrata* in the treatment of equine lameness. [23]

Boswellia serrata versus diseases characterized by inflammation

There is positive evidence that boswellic acids reduce the synovial fluid leucocyte count and lower the elevated serum transaminase levels, as well as erythrocyte sedimentation rates.[21,97] Inflammatory conditions such as rheumatoid arthritis are characterized by a marked increase in the above mentioned parameters. Boswellic acids function as potent anti-inflammatory agent in rheumatic conditions, being especially effective in shrinking inflamed tissues.[32] This action is mediated through a vascular phenomenon. Boswellic acids improve blood supply to the joints and restore the integrity of blood vessels obliterated by spasm. They may also open up collateral blood circulation to provide adequate blood supply to the joints.[10]

A recent review advocates the use of boswellic acids in the treatment of inflammatory diseases of the joints, epidermal lesions (psoriasis), allergic and chronic asthma, inflammatory diseases of the intestines e.g. Crohn's disease, ulcerative colitis, and chronic hepatitis.[24]

Antihyperlipidemic and anti-atherosclerotic action

Alcoholic extracts of *Boswellia serrata* have been shown to significantly lower serum cholesterol, serum triglyceride levels, and deposition of body fat in hyperlipidemic experimental animals.[27,28,31] Studies revealed that boswellic acids are the active principles involved, acting by interference with cholesterol synthesis.[29,30] The treatment with boswellic acids was reported to produce both reversal and prevention of atherosclerosis in experimental animals which were fed high-fat diets.[14]

Hepatoprotective effects

Feeding of the crude ethanolic extract from *Boswellia serrata* to mice was found to exert protective effects against chemical and biological agents causing inflammation and injury to the liver[20], especially hepatitis.

Analgesic, antipyretic and psychopharmacological effects

A non-acidic, oil fraction of the gum resin of *Boswellia serrata* (obtained after the removal of boswellic acids and related compounds) was found to exhibit significant pain relieving effects in rats.[8,33] This analgesic effect was accompanied by marked sedative effects comparable in strength to morphine. It is probable that the analgesic mechanism of the non-acidic fraction is a result of central nervous system sedation.[8,33] The acidic fraction of the gum of *Boswellia serrata*, consisting of boswellic acids, was found to exhibit no analgesic effects in mice and rabbits, but exerted antipyretic (lowered body temperature) action in pyretic rats and rabbits[5,6] (these animals were injected with bacterial toxins (pyrogens) which raise body temperature).

Anti-carcinogenic effects

An alcoholic extract of the resin from *Boswellia serrata* was examined for anti-carcinogenic properties. When tested on mice with Ehrlich ascites carcinoma and S-180 tumor, it inhibited the tumor growth and increased the life span of experimental animals with the carcinoma.[39]

Antimicrobial activity

The essential oil of *Boswellia serrata* was found to exert antifungal properties.[80] Anti-bacterial activity of the oleoresin has also been demonstrated.[37,38]

The broad pharmacological actions of *Boswellia serrata,* and its anti-inflammatory activity in particular, have found important clinical applications as a non steroidal anti-inflammatory drug (NSAID). These properties and related clinical applications of *Boswellia serrata* will be reviewed in the subsequent chapters.

BOSWELLIA SERRATA: BOTANICAL ASPECTS

1. Origin, geographical distribution and varieties

Members of the Natural Order Burseraceae have been used as sources of aromatic compounds and therapeutic agents since ancient times. The genus *Boswellia* Roxb. is constituted by balsamiferous trees that secrete aromatic oleoresins known as olibanum or frankincense. The better known members of this genus (consisting of about ten species of trees and shrubs) are widely distributed in the tropical regions of Asia and Africa. Some of the well known species are listed in the Table[3,46] below:

Species	Geographical Distribution
Boswellia carterii Birdw.	Southern Arabia, East Africa
Boswellia bhaw-Dajiana Birdw.	East Africa
Boswellia serrata Roxb.	India, China
Boswellia frereana Birdw.	East Africa
Boswellia dalzielli Hutch	Nigeria
Boswellia odorata Hutch	Nigeria
Boswellia papyrifera Hutch	East Africa

All these species secrete solid or sticky semi-solid materials, called resins. True resins are soluble in organic solvents, whereas gum resins are only partly soluble in organic solvents. Oleogum resins contain essential oils important in the formulation of perfumes and cosmetics. *Boswellia serrata* produces oleogum resin.

Since ancient times, resins have been important in the preparation of incense, medicines, cosmetics and perfumes. The Egyptians, Hindus, Persians, Israelites, Greeks, Romans, and the Europeans of Queen Victoria's times greatly valued these materials. Olibanum, the resin from the *Boswellia* species has been used as an incense for centuries. However, its major use today is as a fixative in perfumes, soaps, creams, lotions, and detergents.

Boswellia serrata, Boswellia carterii, Boswellia papyrifera and *Boswellia frereana* are reported to have been important commercial sources of olibanum.[93] The resins have been used in India and Africa for the treatment of several diseases and are particularly valued for their anti-inflammatory action.[94] *Boswellia serrata* (var. *serrata* and var. g*labra*) is the only species found in India.[2]

The tree, *Boswellia serrata* , known in Sanskrit as "Gajyabhakshya" (implying eaten by elephants) grows widely on the dry hills throughout India, but particularly in the northwest. Its natural habitat is the mountainous tracts of central India, the Deccan Plateau, Bihar, Orissa, North Gujarat, and Rajasthan. It grows well in the Western Ghats of South India.[108] Two varieties are usually distinguished: var. *glabra* with entire glabrous leaves and var. *serrata* with serrate and pubescent leaves (Fig. 2.1 and 2.2).

It is a deciduous, medium-sized, branching tree with a trunk 3.5 to 4.5m in height and 1.0 to 1.5m in girth. It has a characteristic ash-colored bark peeling off in thin flakes. Young shoots and leaves are pubescent. Leaves are long, opposite, sessile, variable in shape, ovate or lanceolate, crenate-serrate, obtuse, base acute and rounded. Flowers in auxiliary racemes are shorter than the leaves. The calyx is pubescent outside, with broadly triangular, ovate lobes. The petals are long and ovate, the drupe is tringonous, with three pyrenes, each one-seeded.[108] Leaflets are opposite, sessile, 8-15 pairs are present, along with an odd one (the pair at the base of the leaf being much smaller than the others).

Besides the medicinal uses, a major use of the tree is as a source of heart wood for the production of paper pulp in newsprint mills. The collection of gum resin is restricted to a few areas such as the Udaipur, Sirohi, Alwar and Kota districts of Rajasthan, the Morena district of Madhya Pradesh, the Banaskantha district of Gujarat and the Karimnagar district of Andhra Pradesh.

2. Tapping of the oleogum resin

The tapping of the oleogum resin is carried out towards the end of October. A portion of the stem bark about 15 cm to 20 cm wide is scraped away. Transverse incisions are made in the upper and lower portions of the tree trunk and the bark in between is peeled off. The resin exudate resembles Canada Balsam in color and consistency. On exposure to air, the exudate becomes gum like. The oleogum resin is collected after 10-12 days. A single tree yields about 1 kg of resin annually. A tree can be tapped for six to eight years. The brown resinous material is sold un-graded as "salai guggal" or sieved into commercial grades and used in the preparation of incense, fumigants, indigenous perfumes (attars) and medicinal products. The resin in the gum is used in the preparation of varnishes.

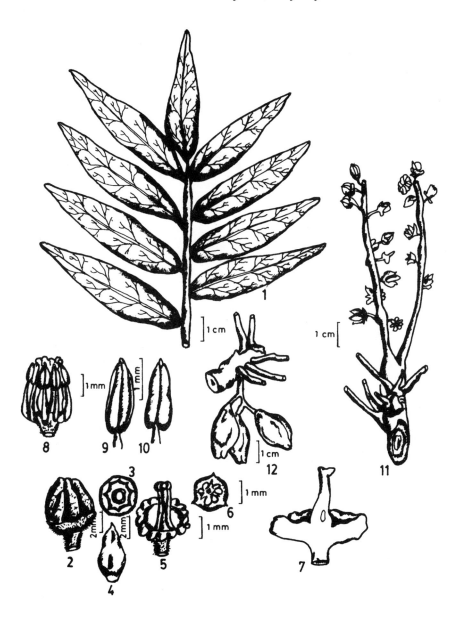

2.1 : *Boswellia serrata var. glabra*
1. Leaf, 2. Flower, 3. Calyx, 4. Petal, 5. Disc with pistil,
6. & 7. Pistil, t.s. & l.s., 8. Stamens and Pistil, 9. & 10. Stamens,
11. Inflorescence, 12. Drupes (Fruits)

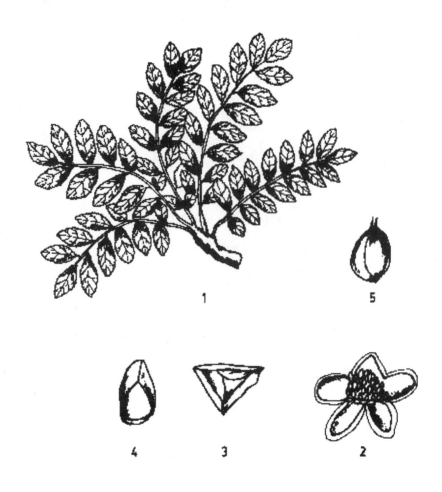

2.2 : *Boswellia serrata var. Serrata*
1.Twig with leaves, 2. Flower, 3. Ovary (c.s), 4. Ovary (l.s) 5. Fruit

BOSWELLIA SERRATA: CHEMISTRY

More than a hundred chemical compounds have been reported in the oleogum resins from the *Boswellia* species. Of these, the two triterpenoid acids, O-acetyl-β-boswellic acid and β-boswellic acid are considered to be the most characteristic.[46]

Composition of the oleogum resin

The physicochemical properties of the oleogum resin of *Boswellia serrata* have been determined by several researchers.[14,50,51] The constituents can be grouped under:

1. Oil
2. Terpenoids
3. Gum

The composition of the oleogum resin as described in the Merck Index (1976)[47] is as follows:

Volatile oil *(pinene, dipentene etc.)*	3-8%
Resins	60%
Polysaccharides	20%
Bassorin	6-8%

Goswami *et al.*[48] and Fowler *et al.*[49] have described methods for the separation of essential oil, terpenoids and gum.

The oily components

The fixed oil is usually pale yellow in color and has a pleasant odor. The yield of essential oil by steam distillation is 16% of the gum resin. The oil is pale yellow in color and has an agreeable balsamic odor.[14] Pearson and Singh[52] reported that α-and β-pinenes were the main constituents. Simonson[53] studied the low boiling fractions of the oil and found α-thujene as the main constituent and α-pinene and α-phellandrene in minor quantities. Roberts[54] studied the high boiling fraction and found the presence of terpenol, methyl chavicol and sesquiterpenes as the major components. Guenther[55] has described the characteristics and uses of the essential oil. The physicochemical characteristics of the oil vary with the source. About 12 individual chemical components have been found in oil from various sources. TLC (Thin Layer Chromatography) and GLC (Gas Liquid Chromatography) studies of the essential oil from the leaves of *Boswellia serrata* have also been performed.[25]

The composition of the oil from the oleogum resin, based on the reported data for individual constituents is as follows:[56]

Alpha Thujene	50%
Alpha Pinene	6.2%
d-limonene	4.5%
p-cymene	14.0%
Cadinene	4%
Geraniol	0.8%
Elemol	1.3%

The terpenoids

The terpenoids of the oleogum resin of *Boswellia serrata* have been investigated for many years. The presence of symbol α-, β-, γ-boswellic acids has been reported by several researchers. These acids were precipitated from the non-volatile fraction of the resin, using barium hydroxide.[57] Preparation of surfactants from some of these acids has also been reported.[60]

These acids (classified as Δ^{12} α-amyrin acids) were first isolated and characterized by Winterstein and Stein[59] who also described the isolation of acetyl-β-boswellic acid on the basis of spectral data and chemical interconversions.

Several reports[60,78,79] indicate that there are four pentacyclic triterpene acids, β-boswellic acid being the major constituent of the mixture. The NMR[61] (Nuclear Magnetic Resonance) and (Mass Spectrometry)[62] data of these triterpenes has also been reported.

A critical examination[63-65] of the non volatile fraction of the resin revealed the presence of the following compounds:
1. Terpene acids
2. Neutral compounds including methyl chavicol, α- and β-amyrins, and a diterpene alcohol, serratol. The phytosterol, β-sitosterol has also been isolated from the bark of *Boswellia serrata*.

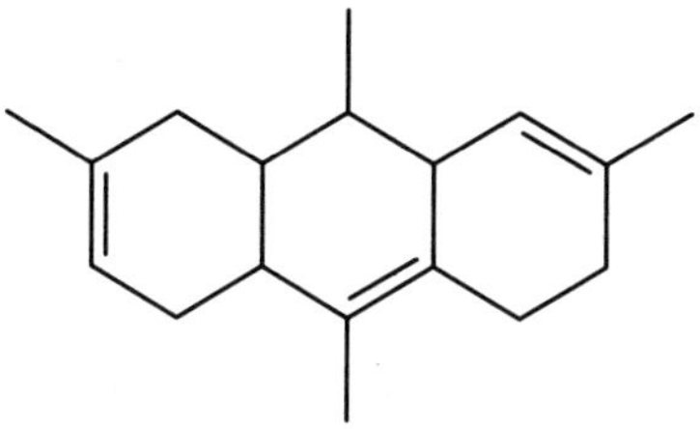

Serratol

3. Four tetracyclic triterpene acids

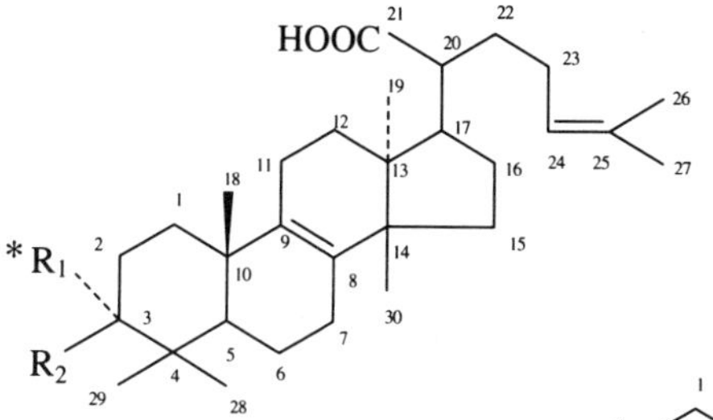

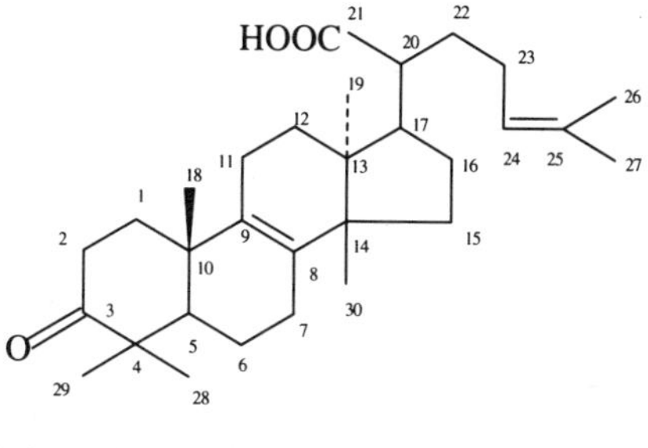

3-ketotirucall-8,24-dien-21-oic acid

*a) R_1 = OH, R_2 = H
(3-α-hydroxytirucall-8,24-dien-21-oic acid)
b) R_1 = H, R_2 = OH
(3-β-hydroxytirucall-8,24-dien-21-oic acid)
c) R_1 = OAc, R_2 = H
(3-α-hydroxytirucall-8,24-dien-21-oic acid)

14

4. Four pentacyclic triterpene acids, including β-boswellic acid and related triterpene acids, now identified as the compounds responsible for the anti-inflammatory and anti-arthritic activities of extracts of *Boswellia serrata*. The structural formulae and physicochemical characteristics of these acids are presented below:[6,15]

General formula

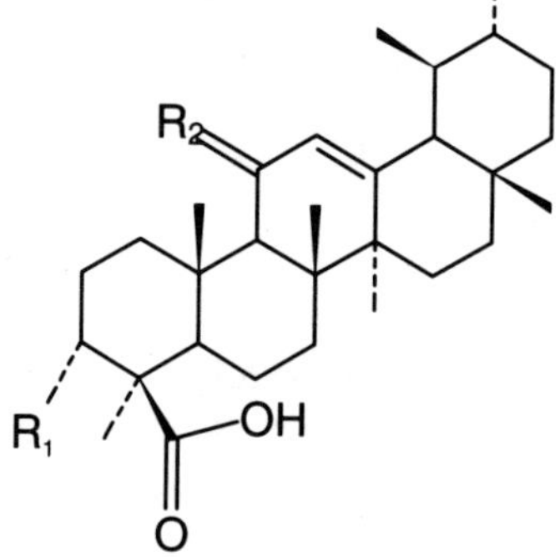

β-boswellic acid

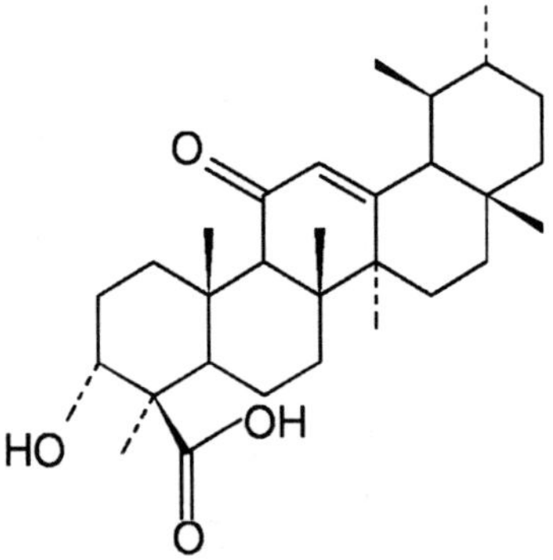

EN: 150515
β-boswellic acid
3α-Hydroxyurs-12-en-23-oic acid
mp 232° C
$C_3OH_{48}O_3$; Mol wt: 456.71

Acetyl-β-boswellic acid

EN: 150516
Acetyl-β-boswellic acid
3-α-Acetoxyurs-12-en-23-oic acid
mp 273° C
$C_{32}H_{50}O_4$; Mol wt: 498.74

11-keto-β-boswellic acid

EN: 150517
11-keto-β-boswellic acid
3-α-Hydroxyurs-12-en-11-keto-23-oic acid
mp 270° C.
$C_{30}H_{46}O_4$; Mol wt: 470.69

16

Acetyl-11-keto-β-boswellic acid

EN: 150518
Acetyl-11-keto-β-boswellic acid
3-α-Acetoxyurs-12-en-11-keto-23-oic acid
mp 251° C
$C_{32}H_{48}O_5$; Mol wt: 512.73

The pharmacologically active salts of these boswellic acids include sodium, potassium, calcium and ammonium.

Based on β-boswellic acid, structural modifications were carried out by researchers to obtain compounds with enhanced anti-inflammatory activity.[67,83,103] β-boswellic acid was condensed with different substituted aromatic aldehydes to obtain a number of compounds, some of which showed anti-inflammatory activity. The dimethoxy, m-hydroxy and p-methoxy analogs showed moderate anti-inflammatory activity in rats. The 3-pyrazole derivative synthesized from β-boswellic acid was found to be 50% more active than the total boswellic acids.[83]

Ethyl esters, diformate and dibenzoate derivatives of β-boswellic acids have also been synthesized. Some of these are represented below:

A: α-boswellic acid derivatives   B: β-boswellic acid derivatives

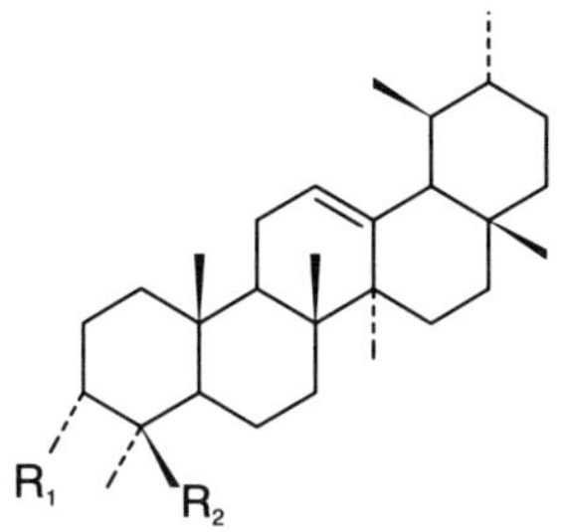

R_1	R_2	R_1	R_2
OH	CO_2Et	OH	CO_2Et
OAc	CO_2Et	OAc	CO_2Et
OBz	CO_2Me	HCO_2	CH_2O_2CH
		$PhCO_2$	CH_2O_2CPh

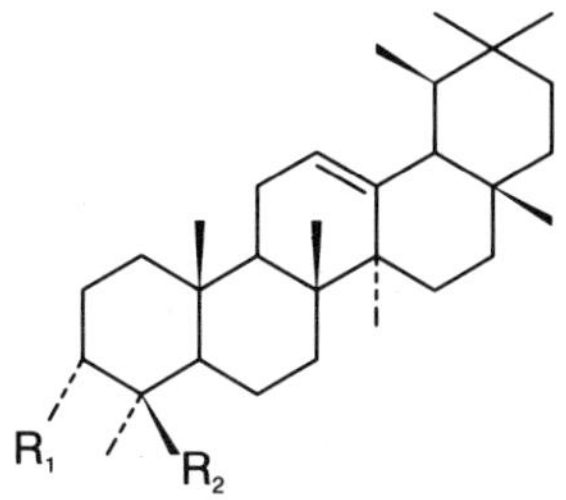

As pharmacologically effective derivatives of boswellic acids, lower alkyl esters, which are obtained by the esterification of the carboxyl group with a C_1 to C_6 alcohol (preferably the methyl ester), or esters which are obtained by the esterification of the hydroxyl group with carbonic acid, can be used.[68]

The derivatives preferred for anti-inflammatory and anti-arthritic action include β-boswellic acid acetate, β-boswellic acid formate, β-boswellic acid methyl ester, acetyl-β-boswellic acid, acetyl-11-keto-β-boswellic acid and 11-keto-β-boswellic acid. Of these, 11-keto-β-boswellic acid was found to be the most potent antiphlogistic and antirheumatic agent, due to the presence of both the carboxylic functional group and an 11-keto functional group.[69]

Isolation and detection of boswellic acids

The method described by Winterstein and Stein[59] has been used in the past for the isolation and purification of boswellic acids.

The current method[6] involves the precipitation of the boswellic acids as barium salt, which is converted to a mixed anhydride of acetic acid and boswellic acids. Acetyl-boswellic acid is obtained from this mixture by boiling with ether-methyl alcohol, followed by crystallization from methyl alcohol. Aliquots of acetyl-boswellic acid are then saponified with methanolic KOH, crystallized from methyl alcohol, washed with water and dried under vacuum at 140° C. The α- and β- isomers of boswellic acid and the 11-keto-β-boswellic acid, as well as the acetyl-boswellic acid derivatives, are then isolated by HPLC (High Pressure Liquid Chromatography). The isolated boswellic acids and their derivatives are characterized by MS (Mass Spectrometry), IR (Infra red Spectrometry), UV (Ultra-violet Spectrometry) and ¹H-NMR (Hydrogen Nuclear Magnetic Resonance).

The scheme followed by a recent group of researchers[6] for the isolation of boswellic acids is outlined below:

SCHEME FOR THE ISOLATION OF BOSWELLIC ACIDS

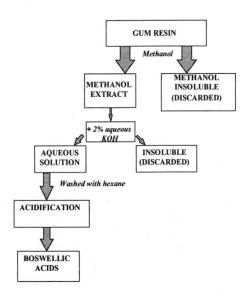

19

In the commercial process of the isolation of boswellic acids, the gum resin is defatted at room temperature. This defattening process removes about 40% of the fat and essential oil. After defatting, the residue is subjected to percolation, selective extraction, fractionation and concentration under vacuum to give a cream colored powder. The yield at this point is about 12% of the starting material. This proprietary extraction process gives a mixture of triterpene pentacyclic acid derivatives of boswellic acids, the major single component of which is β-boswellic acid at approximately 30%.[16]

A non-aqueous titrimetric method was developed for the estimation of total triterpene acids in the oleogum resin of *Boswellia serrata*. This method was based on β-boswellic acid, which constitutes more than 30% of the total triterpene acids. The other constituents in the total triterpene acids are 11-keto-β-boswellic acid and acetyl-11-keto-β-boswellic acid. The authors of this study estimated triterpene acids alone or in combination of two, using functional group analysis. The functional groups analysed were acetyl and hydroxy groups and the keto group at the 11-position.[71]

The triterpene acids are conveniently detected by rapid and sensitive chromatographic methods. GC/MS distinguishes β-boswellic acid and O-acetyl-β-boswellic from other triterpenoic acids. A TLC procedure helps to separate these two acids, the hydroxy acids being strongly retained by the polar silica gel used in this method.

The gum fraction

Samples of the oleogum resin analysed by the Imperial Institute,[72] London (1919) showed the following composition:

Moisture	10-11%
Volatile Oil	8-9%
Resin	55-57 %
Gum	20-23%
Insoluble matter	4-5%

Malandkar[73] hydrolysed the gum by heating it with 3% sulphuric acid for eight hours and identified the resultant sugars as arabinose, xylose and galactose. However, recent studies[74,76,77] identified glucose, arabinose, rhamnose, galactose, fructose, glucuronic acid and idose as hydrolytic products. The proximate composition of the gum was determined as:[74]

Moisture	13.02%
Total ash	1.9%
Acid-insoluble ash	0.2%

These values are the distinguishing characteristics for salai gum as they differ considerably from those of acacia and tragacanth gums.

The gum is not a good suspending agent, but its emulsifying properties are superior to acacia gum. Tablets prepared using 9% salai gum mucilage were comparable to those prepared by using 5% acacia gum mucilage.[74] The gum contains oxidizing and diastatic enzymes. The oxidase system consists of peroxidase, an oxygenase, and a substance giving reactions characteristic of the catechol group.[49]

BOSWELLIA SERRATA: PHARMACOLOGY, PRECLINICAL AND CLINICAL STUDIES

Salai guggal, the oleogum resin of the *Boswellia serrata* has been used in Ayurvedic medical practice to treat a variety of diseases including arthritis, inflammatory conditions, liver diseases, neurologic disorders, obesity, respiratory tract disorders and skin infections. The anti-inflammatory and anti-arthritic effects of salai guggal (in the form of defatted alcoholic extracts) were investigated by researchers at the Regional Research Laboratory, Jammu, India to validate the effectiveness of extracts from the oleogum of *Boswellia serrata* in the treatment of inflammatory disorders. These studies were undertaken as part of the search for useful anti-inflammatory drugs of natural origin which do not harbor the side effects of the currently used non-steroidal anti-inflammatory drugs (NSAID's) and steroidal anti-inflammatory agents, such as GI bleeding, ulceration and/or perforation.

Other groups of researchers found that the non-acidic portion of the extracts exhibited analgesic and psychopharamcological effects.[8] In addition, hepatoprotective properties,[20] antimicrobial effects and anti-carcinogenic properties of Salai guggal have also been identified, confirming the diverse biological properties of the oleogum resin of *Boswellia serrata.*[14,15]

Several groups of researchers studied the effects of the active principles from the resin, the boswellic acids, as well as unfractionated extracts of the resin. Pre-clinical studies in rodents, rabbits and primates as well as clinical studies on human volunteers have been performed and the standardized extract has been recognized as safe for human and veterinary use.[5,6]

1. Pre-clinical studies for therapeutic effectiveness of the the oleogum resin of *Boswellia serrata:*

I. Analgesic and psychopharmacological effects

Preliminary studies on rats revealed that the gum resin of *Boswellia serrata* possessed a marked analgesic property and sedative effect.[33] This psychopharmacological action was further attributed to the non-acidic fraction of the gum resin, isolated by removing the fraction containing boswellic acids and related compounds, from the destructively-distilled gum resin.[8] The non-acidic fraction of Salai guggal produced central nervous system (CNS) sedation in rats, characterized by less spontaneous activity and dropping of the palpebrae. This latter sign is a typical sign of sleepiness, which is a result of sedation, and is experienced by those who use barbiturate containing sleeping pills. The sedative effect of the non-acidic fraction of Salai guggal was further evidenced by the potentiation of sleepiness induced experimentally with barbiturates.

The non-acidic fraction produced a significant pain relieving analgesic effect in rats. This analgesic effect was evident within 30 minutes of the compound administration, and lasted for about two hours. The pain relieving action coincided with the sedative effect, and was attributed to the sedation.

The psychopharmacological mechanism of this non-acidic fraction has been compared to that of morphine. The analgesic effect was similar in strength for both drugs, and it was diminished when the animal was pre-treated with a small dose of morphine and then exposed to an analgesic dose of the non-acidic compound. This experiment shows that morphine and the Salai guggal fraction share similar receptors in the CNS. It has also been found that the biological effect of the non-acidic fraction was partially neutralized by the drug nalorphine, which is an antagonist to morphine and used in the treatment of morphine overdose. This observation indicates that the non-acidic fraction of Salai guggal shares the CNS receptors with those two drugs.

Further investigation of the psychopharmacological properties of Salai guggal showed that the non-acidic fraction does not significantly reduce the anxiety states in rats subjected to trained avoidance response. The fraction was compared with a standard anti-anxiety drug chlorpromazine (which produced tranquilizing effects in conditioned animals). The non-acidic fraction of the oleogum resin was therefore found to have central nervous system depressant and analgesic actions, but was less effective in reducing the anxiety levels in rats, as compared to the standard anti-anxiety drug chlorpromazine.

A comparison of the analgesic effects of the non-acidic extract with morphine hydrochloride, is presented in (Fig. 3.1).

Analgesic effect of the non-acidic fraction of Salai guggal

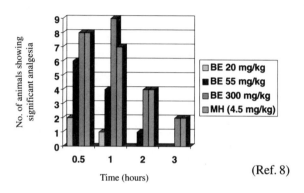

(Ref. 8)

BE : non-acidic fraction of Salai guggal extract
MH : morphine hydrochloride

Fig. 3.1

The non-acidic fraction separated from the gum resin of *Boswellia serrata* was found to have marked hypotensive and cardiovascular effects in dogs, probably due to peripheral blood vessel dilation. A dose of 300 mg/kg of the fraction lowered blood pressure (by 50-60 mm of mercury) and the effect lasted for 15-20 minutes. A dose of 15 mg/kg caused a transient decrease in the heart rate of dogs[95].

The other effects of the non-acidic fraction tested in experimental models included an increase in the heart rate of a frog heart,[95] and a decrease in gastrointestinal contractions produced by acetylcholine and histamine in guinea pig and rabbit ileum.[95]

II. Anti-inflammatory and anti-arthritic effects

The major use of *Boswellia* oleogum resin is in the treatment of inflammatory diseases. Inflammation results from a complex series of actions and/or reactions triggered by the body's immunological response to tissue damage. Many diseases, as well as physical trauma, induce inflammatory reactions. These reactions, although necessary to start the healing process, often escape the regulatory control mechanisms and may lead to chronic conditions such as arthritis or hepatitis.

Mechanism of anti-inflammatory action:

Arachidonic acid is an essential fatty acid synthesized in the body. It is a precursor for the biosynthesis of important hormone-like substances such as prostaglandins of the 2 series, thromboxanes and leukotrienes, which play major roles in the process of inflammation. Arachidonic acid is mobilized from the phospholipids in the body by the action of the enzyme phospholipase. It can be further converted through the enzyme cyclooxygenase to pro-inflammatory prostaglandins (PG). By the action of another enzyme, lipoxygenase, it is converted to hydroxyeicosatetraenoic acids (HETE) and leukotrienes (LT). Some of the prostaglandins like PGE_2 and PGI_2 dilate the blood vessels, while certain leukotrienes such as $LTB_{,4} LTC_{,4}$ and LTD_4 increase vessel permeability resulting in tissue swelling and tenderness, which characterizes inflammation. Increased levels of some prostaglandins like PGE_2 produce redness, swelling and pain in the inflamed part of the body, which are recognized as signs of inflammation. The contribution of other factors like prostaglandin TXA_2 and leukotrienes LTC_4 and LTD_4 to inflammation is by cutting off the blood and nutrient supply to the tissue.[17]

Steroidal drugs like cortisone, and non-steroidal anti-inflammatory drugs (NSAID's) like ibuprofen, aspirin, phenylbutazone and indomethacin are used in clinical practice to subdue inflammation. Some of the anti-inflammatory drugs inhibit the lipoxygenase pathway, some arrest the cyclooxygenase pathway, resulting in differing potency and clinical applications as anti-inflammatory agents. Unfortunately, most of the anti-inflammatory drugs, particularly steroids, besides being effective, can also produce dangerous side effects such as gastric irritation leading to ulceration, bleeding, altered metabolism, lowered immune system response, and deterioration of the cardiovascular system.

Studies on the anti-inflammatory mechanism of the pentacyclic triterpene acids fraction of the oleogum resin of *Boswellia serrata* indicate that the primary site of action of this compound is inhibition of the 5-lipoxygenase enzyme, preventing the formation of inflammatory leukotrienes.[17,75]

The role of the *Boswellia serrata* extract in the arachidonic acid cascade is indicated in (Fig. 3.2).

MECHANISM OF ANTI-INFLAMMATORY ACTION OF BOSWELLIC ACIDS (acid fraction of the gum resin)

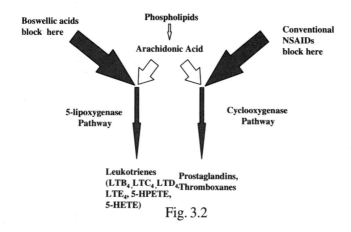

Fig. 3.2

Boswellic acids have been found to inhibit leukotriene synthesis via inhibition of 5-lipoxygenase, but did not affect the cyclooxygenase activities and therefore synthesis of prostaglandins. This data suggests that boswellic acids are specific inhibitors of leukotriene synthesis acting either by interlocking with the 5-lipoxygenase enzyme or by blocking its mobility. The mechanism of boswellic acids and their derivatives were compared with that of nordihydroguaiaretic acid. Both compounds inhibited enzyme 5-lipoxygenase at comparable dose levels. Unlike nordihydroguaiaretic acid, however, boswellic acids did not impair the other two enzymes metabolizing arachidonic acid, cyclooxygenase and 12-lipoxygenase.[15]

Inhibition of the 5-lipoxygenase pathway by various isomers and derivatives constituting the mixture of boswellic acids of *Boswellia serrata* have been studied. It has been found that the acetyl-11-keto-β-boswellic acid provides the best inhibitory action[24], as shown in (Fig. 3.3).

**Biological activity of various derivatives of Boswellic Acids (BA)
in inhibiting the enzyme 5-Lipoxygenase**

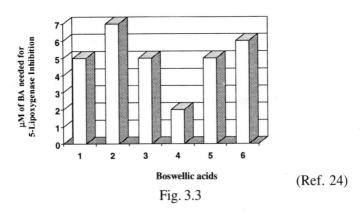

Boswellic acids   (Ref. 24)

Fig. 3.3

Boswellic acids used in this study: (Ref. 24)
1. β-boswellic Acid
2. Acetyl-β-boswellic acid
3. 11-Keto-β-boswellic acid
4. Acetyl-11-Keto-β-boswellic acid
5. β-boswellic acid
6. Acetyl-β-boswellic acid

The dose dependent inhibition of leukotriene LTB_4 formation by an ethanolic extract of Salai guggal is shown in (Figs. 3.4 and 3.5.)[75] The IC_{50} (effective inhibitory concentration of tested substance) value for this extract was 30 mg/ml.

Concentration dependent inhibition of LTB_4 formation by Salai Guggal Ethanolic Extract (SGEE)

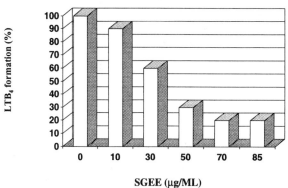

SGEE (μg/ML)

Inhibition of 5-Lipoxygenase (5-LOx) dependent leukotrienes formation by Salai Guggal Ethanolic Extract (SGEE)

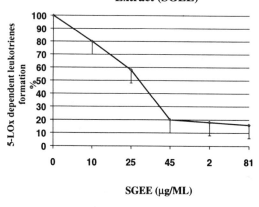

SGEE (μg/ML)

(Ref. 75)

Fig. 3.5

The extract of Salai guggal inhibited leukotriene LTB_4 formation in the suspension of white blood cells derived from a rat. White blood cells are involved in the inflammatory reactions in many ways, and are the site of manufacture for inflammatory substances like leukotrienes. This experiment shows the relationship between inhibition of the enzyme 5-lipoxygenase and the anti-inflammatory activity of the extracts of *Boswellia serrata* gum.

Additional proof for a well defined mechanism of the anti-inflammatory action of boswellic acids came from *in vivo* experiments. Boswellic acids failed to prolong the gestation period in, or delay the onset time of castor oil induced diarrhea in rats[5]. Both outcomes depend upon the availability of prostaglandins. This experiment demonstrates that boswellic acids do not inhibit prostaglandin synthesis, and selectively inhibit the synthesis of leukotrienes.[5]

Further studies on rats revealed that the anti-inflammatory activity of ethanolic extracts of *Boswellia* gum resin is not mediated through the pituitary-adrenal mechanism. This was shown in the experiment in rats with surgically removed adrenal glands; adrenal glands among other functions are responsible for the production of anti-inflammatory hormones, such as corticosteroids. Boswellic acids administered to rats without adrenal glands were still able to prevent inflammatory swelling produced experimentally by a chemical irritant, carrageenan. Therefore, boswellic acids do not require the release of anti-inflammatory corticosteroids by the body for their anti-inflammatory mechanism (Fig. 3.6).

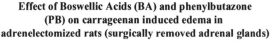

Effect of Boswellic Acids (BA) and phenylbutazone (PB) on carrageenan induced edema in adrenelectomized rats (surgically removed adrenal glands)

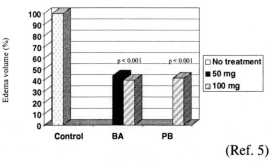

(Ref. 5)

Fig. 3.6

Anti-inflammatory activity:

Boswellic acids were tested in experimentally induced inflammation in rats[5]. The inflammation was produced by chemicals which act as irritants and cause tissue swelling, i.e. carrageenan, croton oil and dextran. The degree of inflammation was measured by the extent of tissue swelling, and the effectiveness of the treatment was estimated by prevention or inhibition of the tissue swelling. Tissue swelling is one of the first inflammatory reactions of the injured or irritated tissue in the process of inflammation. Even a minor cut produces swelling, and this sign of inflammation has been experienced by virtually everyone.

Administration of boswellic acids at a dose range of 50-200 mg/kg orally, produced significant inhibition in the carrageenan induced swelling in the paw of rats. The treatment produced 39-75% inhibition in swelling, compared to 47% inhibition produced by the administration of the known non-steroidal anti-inflammatory drug phenylbutazone (Fig. 3.7).

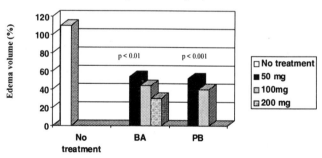

Inhibition of carrageenan induced edema in rats by oral administration of Boswellic Acids (BA) and Phenylbutazone (PB)

(Ref. 5)

Fig. 3.7

30

Boswellic acids isolated from Salai guggal consist of a mixture of pentacyclic triterpene acids. Derivatives of boswellic acids were tested for their individual anti-inflammatory potential. The results of these experiments are shown in (Fig. 3.8). The pyrazol derivative of β-boswellic acid showed the strongest anti-inflammatory action, while 3-acetyl-β-boswellic acid had the least activity.

Anti-inflammatory activity of six boswellic acid derivatives (BA) on carrageenan induced edema in rats

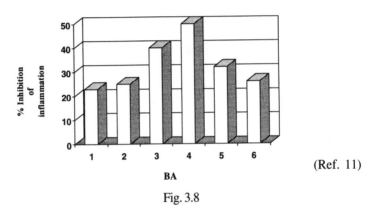

(Ref. 11)

Fig. 3.8

Boswellic acids (BA) used in this study:

1. 3-Acetyl-β-boswellic acid
2. Acetyl-11-keto-β-boswellic acid
3. 12-keto-β-boswellic acid
4. Pyrazol derivative of β-boswellic acid
5. 3-keto-β-boswellic acid ester
6. 2-bromo-β-boswellic acid ester

An ointment containing boswellic acids was effective in reducing croton oil induced and carrageenan induced inflammation in rats, when applied topically at various dose levels.[36] A comparison of the anti-inflammatory activity of the boswellic acids ointment versus the phenylbutazone ointment is presented in (Fig. 3.9).

**Effect of Boswellic Acids (BA) and Phenylbutazone (PB),
topically applied, against
carrageenan induced edema in rats**

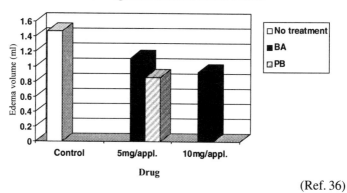

(Ref. 36)

Fig. 3.9

The comparison of various doses of boswellic acids and phenylbutazone in subduing the inflammatory swelling, as depicted in (Fig. 3.9) (topical use) and (Fig. 3.10) (oral administration) showed that boswellic acids (BA) are effective at higher concentrations than is phenylbutazone (PB).

**Effect of boswellic acids (orally administered) against
carrageenan induced edema in rats**

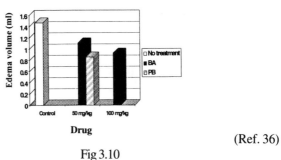

(Ref. 36)

Fig 3.10

Anti-inflammatory vs. anti-oxidant properties of boswellic acids

Boswellic acids, as Boswellin,® applied topically to a mouse ear, exposed to a known chemical irritant and cancer promoting agent TPA, prevented inflammatory reaction of the ear tissue, as measured by the degree of ear swelling (Fig. 3.11). This therapeutic, anti-inflammatory effect of boswellic acids was comparable in strength to that produced by a topical application of curcuminoids. Curcuminoids, as Curcumin C^3 Complex,™ is a well known anti-inflammatory and anti-oxidant natural compound isolated from the roots of turmeric, *Curcuma longa.*

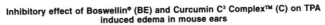

Inhibitory effect of Boswellin® (BE) and Curcumin C^3 Complex™ (C) on TPA induced edema in mouse ears

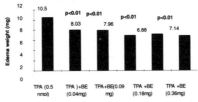

Fig. 3.11

In a similarly designed experiment, Boswellin® prevented another inflammatory compound, arachidonic acid, from causing swelling of the mouse ear. This anti-inflammatory effect was also comparable in strength to the anti-inflammatory effect obtained with Curcumin C^3 Complex™ (Fig.3.12).

Inhibitory effect of Boswellin® (BE) and Curcumin C^3 Complex™ (C) on arachidonic acid (AA) induced edema in mouse ears

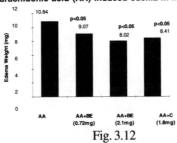

Fig. 3.12

The results of the anti-inflammatory testing of Boswellin,® particularly those showing inhibition of the TPA induced inflammation, are of particular importance, since recent studies have indicated that several compounds that possess anti-inflammatory activity inhibit TPA induced experimental tumors.

One of the well documented natural compounds which inhibits TPA induced tumors is curcumin. The mechanism of TPA in tumor promotion is most likely due to its stimulation of rapid cell growth, as occurs in tumor development, due to induction of the enzyme ornithine decarboxylase (ODC). This is one of the key enzymes which regulate the biosynthesis of polyamines such as putrescine, spermidine and spermine derived from the amino acid ornithine. Several classes of anti-tumor agents inhibit the induction of the ODC this way, presumably, exerting their anti-tumor activity[110]. Interestingly, compounds which inhibit polyamine synthesis may also exert a strong anti-parasitic effect. One of the inhibitors has been shown to be effective in cases with African sleeping sickness, and to be effective against *Pneumocystis carinii*, an opportunistic parasite most commonly encountered in patients with AIDS.[109] A further understanding of the mechanisms of polyamine biosynthesis and inhibition with compounds like boswellic acids and curcumin may result in better treatments of various proliferative and parasitic disorders affecting man.

The probable mode of action of boswellic acids in the inhibition of ODC induction is indicated in (Fig. 3.12(a)).

ROLE OF ORNITHINE DECARBOXYLASE (ODC) IN POLYAMINE METABOLISM
POSSIBLE ROLE OF BOSWELLIC ACIDS (BA) IN PREVENTING TPA INDUCTION OF ODC ACTIVITY

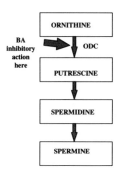

Fig. 3.12(a)

Despite the similarities between the biological mechanisms of boswellic acids and curcumin, these two compounds differ significantly in their antioxidant properties. Boswellic acids, as Boswellin,® tested by the rancimat method, which evaluates protection against fatty acid oxidation, and in the free radical scavenging method showed no anti-oxidant activity, as compared to the untreated control and control treated with curcuminoids. Curcuminoids, as Curcumin C^3 Complex™ displayed significant anti-oxidant activity in both of the tested methods viz., in the prevention and scavenging of free radicals (Fig. 3.13 and 3.14).

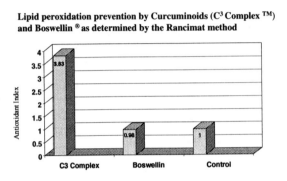

Lipid peroxidation prevention by Curcuminoids (C^3 Complex ™) and Boswellin ® as determined by the Rancimat method

Note: The higher the anti-oxidant index, the better the anti-oxidant action.

Fig. 3.13

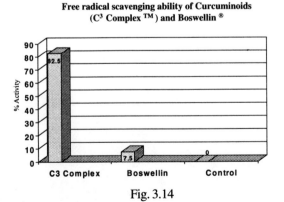

Free radical scavenging ability of Curcuminoids (C^3 Complex ™) and Boswellin ®

Fig. 3.14

The comparative evaluation of boswellic acids and curcumin for their antioxidant and anti-inflammatory properties indicates that the two compounds may afford an anti-inflammatory action with two distinct mechanisms, which may or may not involve the antioxidant mechanism. This observation is particularly interesting in view of a commonly held opinion among researchers that a compound with anti-inflammatory properties is also an antioxidant, and that the anti-oxidant mechanism is responsible for the anti-inflammatory action.

Anti-arthritic activity:

Boswellic acids were evaluated in various experimental models of arthritis. The compound produced clinically significant anti-arthritic activity in cases of complete Freund adjuvant induced arthritis in rats, formaldehyde induced arthritis in rats, gouty arthritis in dogs and bovine serum albumin induced arthritis in rabbits[81]. In addition, signs of inflammatory lesions associated with arthritis, like hemorrhagic patches on the ears, nodules on the tail and swelling of limbs in affected animals, were alleviated with the administration of boswellic acids.

Besides improving the clinical condition of the animals, boswellic acids positively affected various laboratory indices of the arthritis associated inflammation. They inhibited white blood cell migration, or movement across the blood vessels to the inflamed tissue, and reduced the serum levels of transaminase enzymes in carrageenan treated rats[21]. Both parameters, mobility of white blood cells and the increased transaminase enzyme activity, are the cause and/or result of the inflammatory process.

Boswellic acids exhibited an inhibitory effect on the expression of the immune response which plays an important role in some forms of inflammatory reactions, like for example those associated with arthritis. The prominent part of the anti-inflammatory and thus anti-arthritic mechanism of boswellic acids is the inhibition of arachidonic acid metabolism and also the formation of leukotrienes[75].

The wasting of body tissues due to inflammation induced destruction and catabolic reactions is a prominent clinical sign of chronic inflammatory processes (such as those associated with arthritis). Anabolic and catabolic reactions are the two ongoing metabolic processes in the body. Anabolic reactions lead to incorporation of nutrient materials into the body tissues and organs; catabolic reactions use up stored materials such as sugars, fats and protein, to yield energy. When in balance, both types of reactions serve the vital role of energy supply and storage in the body.

Predominantly, catabolic reactions in chronic arthritis inflammation cause loss of body weight[5] and degradation of connective tissue[34]. As a result, the base substance of the building tissue, chemically known as glycosaminoglycans, is being "chipped" away leading to the destruction of the supportive tissues forming the joint. This leads to continuously worsening joint disfigurement and limited mobility. Subsequent to this destruction, the urinary excretion of connective tissue metabolites in animals with adjuvant arthritis is usually higher than that in normal animals.

Treatment with boswellic acids and extracts of Salai guggal in rats suffering from adjuvant arthritis, reduced the "chipping" of glycosaminoglycans, thus preventing the basic mechanism leading to clinical deterioration in arthritis patients[34].

Furthermore, treatment with boswellic acids reduced various tissue destructive enzymes, namely lysosomal glycohydrolases, including beta-glucuronidase, beta-N-acetylglucosaminidase, cathepsin B1, cathepsin B2 and cathepsin D. Ultimately, experimental treatment with boswellic acids diminished urinary excretion of the connective tissue metabolites, i.e. hexosamine, hydroxyproline and uronic acid[34].

The overall mechanism of boswellic acids in connective tissue metabolism is depicted in (Fig. 3.15).

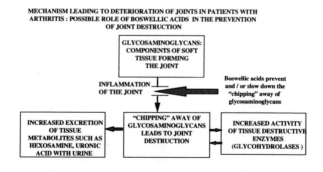

Fig. 3.15

As demonstrated by (Figs. 3.16-3.19), boswellic acids administered to rats orally in the dose range of 50-200 mg/kg of body weight
displayed anti-arthritic activity comparable to that of phenylbutazone.
As a result of the treatment with either boswellic acids or phenylbutazone,
the body weight loss associated with arthritis was significantly prevented.[5,7]
(Fig. 3.17). Also, levels of liver enzymes, elevated due to systemic in-
volvement in the course of arthritis, were significantly lowered in animals receiving boswellic acids or phenylbutazone (Fig. 3.19).

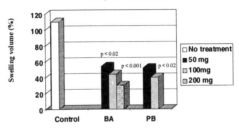

(Ref. 5)

Fig. 3.16

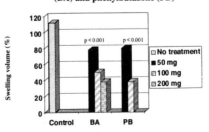

(Ref. 5)

Fig. 3.17

**Dose dependent prevention of inflammation induced
loss of body weight in rats by Boswellic Acids (BA)
and Phenylbutazone (PB)**

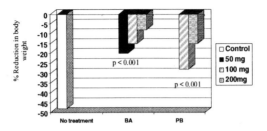

(Ref. 5)

Fig. 3.18

**Effect of Boswellic Acids (BA) and phenylbutazone (PB)
on liver enzyme levels in formaldehyde induced Arthritis**

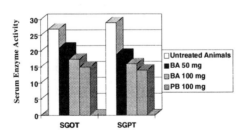

(Ref. 5)

Fig. 3.19

Being effective anti-inflammatory compounds, boswellic acids were
further studied as potential immunological response modifiers. The immune response of the organism plays a critical role in two ways: by
enhancing and subduing the inflammatory process.

Role of the immune system in our body, an overview:

The immune system functions to defend the integrity of vital body functions. The primary factors of this immunological defense are blood cells called macrophages and lymphocytes. These cells are often referred to as white blood cells, to distinguish them from red blood and tissue cells which are involved mostly in the delivery of oxygen to the body. The action of macrophages throughout the body is widespread. They are the first foot soldiers to rush to the site of infection, where their immediate action is to "eat" (phagocytize) foreign particles (antigens) like bacteria, viruses, and parasites. This process, technically called phagocytosis, carries out another valuable function: determination of exactly what kind of antigen has been phagocytized, or eaten. This process is called antigen recognition.

The information obtained by the macrophages is passed on to other, more sophisticated cells. These are the lymphocytes, which function like generals and are responsible for the body's final line of defense. Lymphocytes are divided into two main groups: B-lymphocytes and T-lymphocytes. B-lymphocytes are responsible for the body's humoral response through antibodies. They produce and secrete these antibodies, which are soluble substances, to fight any infection not completely controlled by the macrophages. To make it more clearly understood, antibodies can be thought of as anti-antigens. At the first encounter with an antigen, B-lymphocytes produce a so called primary response antibody, with subsequent encounters, they produce a secondary response antibody.

On the other hand, T-lymphocytes are responsible for cellular defense mechanisms, the most sophisticated activity of the immune system. The cellular defense mechanism triggers more than just antibodies to be sent out like torpedoes against antigens. In addition to directly destroying antigens, these T-lymphocytes also function as the chief of staff (who presides over a group of generals) to regulate the entire immune response.

Balanced interaction between the three immunological cells, tissue and blood macrophages, B-lymphocytes and T-lymphocytes represent self regulation of the immune system. The intact self regulation is regarded as the most crucial aspect of the immune response to any outside challenge to the body either by a microorganism or any form of stress, be it be psychological, physical, chemical or biological.

In diseases with chronic inflammatory conditions, exemplified by arthritis, the self regulation of the immune response is seriously disturbed, resulting in the uncontrolled hostile reaction of the immune system towards its own host. Basically, the immune system is treating tissues of the body as "antigens" and "eating" them away. This is a self perpetuating mechanism which is called "auto-immune disease". Rheumatoid arthritis is an example of an auto-immune disease. The logical intervention for auto-immune diseases is to subdue the immune response. This is accomplished by anti-inflammatory drugs which suppress any and all steps of the immune response, as shown in (Fig. 3.20):

SCHEME OF THE IMMUNE RESPONSE : POSSIBLE SITES OF ACTION OF BOSWELLIC ACIDS

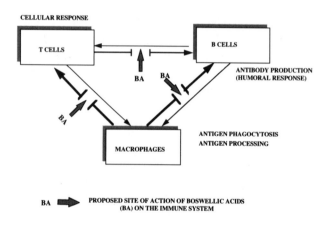

Fig. 3.20

Effect of boswellic acids on humoral and cell mediated immunity

Oral administration of boswellic acids in a dose range 50-200 mg/kg of body weight caused dose related inhibition of antibody production in response to the antigen challenge in experimental animals. The primary antibody response was affected more significantly by boswellic acids than was the secondary antibody response. Boswellic acids were also effective in inhibiting the cell mediated response [105] subsequent to the antigen challenge.

Effect of boswellic acids on the inflammatory cells and proteins arthritis:

As previously mentioned, immune mechanisms play an important role in the development of arthritis. In fact arthritis can be induced in experimental conditions by injecting a variety of antigens into the joints of animals. A chronic arthritis resembling human rheumatoid arthritis was produced in rabbits by means of systemic and local injection with an antigen in the form of a bovine serum albumin (BSA).[106] The severity of experimentally induced arthritis was evaluated by analyzing the content of the inflammatory fluid obtained from the arthritis affected joint. This fluid contains inflammatory white blood cells and certain proteins which increase due to inflammation.

In one experiment, quantitative and qualitative evaluation of the content of white blood cells in the inflammatory fluid of the joint was done with and without the administration of boswellic acids. Also the patterns of proteins (inflammation markers) in the inflammatory fluids were compared. As a result of boswellic acids administration, the number of white blood cells was significantly reduced in the inflammatory fluid, and the pattern of the proteins was also altered.

A comparison of total white blood cell (WBC) counts of the knee joint fluid obtained at various intervals of the BSA-induced arthritis in animals treated with 25, 50 and 100 mg/kg body weight of boswellic acids is shown in (Fig. 3.21).[81]

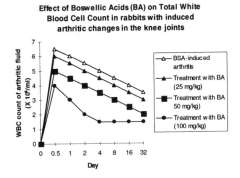

Fig. 3.21

42

Anti-complimentary activity of boswellic acids:

Boswellic acids were found to possess anti-complementary activity,[97] which may partially explain the anti-inflammatory mechanism of the compounds. Complement is a complex of proteins in the body which is an integral part of the immune system and the immune response. The complement system responds to, and may be activated by foreign antigen overload or an infection with various microorganisms. Its activation is part of the defense reaction of the body, which leads to the elimination of the infected cells and elimination of infectious agents like bacteria or viruses. Unfortunately, this self defense mechanism often results in a severe inflammatory reaction which per se is detrimental to the recovery of the primary disease.

The complement system has also been implicated in chronic inflammation and destruction of the joints and joint tissues in various forms of arthritis. In a way, the complement system mediates inflammation by interacting with the antibody which is bound to an antigen. This reaction between the three elements, antigen, antibody and complement, results in the destruction of the cells. The active substances released contribute to the inflammation by increasing edema formation, stimulating inflammatory white blood cells, releasing lysosomal enzymes that damage the tissue and by stimulating the production of inflammatory arachidonic acid metabolites, viz., prostaglandins and leukotrienes.

Boswellic.acids were found to inhibit *in vitro* a complement dependent destruction of sheep erythrocytes.[21] An *in vivo* correlation between boswellic acids anti-inflammatory and anti-complemantary activities has also been documented.[97] These results indicate that boswellic acids mediate the anti-inflammatory activity through the direct inhibition of complement activation, or indirectly by inhibiting complement mediated inflammatory reactions (Fig. 3.22).

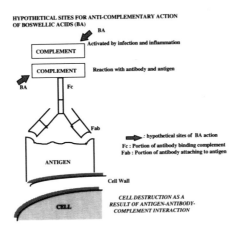

HYPOTHETICAL SITES FOR ANTI-COMPLEMENTARY ACTION
OF BOSWELLIC ACIDS (BA)

Fig. 3.22

III. Anti-hepatotoxic effects of *Boswellia serrata*

Recent studies showed that the administration of endotoxins to rats led to an increase in leukotriene secretion into the bile. Here leukotriene synthesis inhibitors exerted protective effects against endotoxin induced, as well as chemical galactosamine induced, injury to the liver in mice.[20] One way to evaluate damage to the liver caused by toxic substances like galactosamine/endotoxins is by evaluating levels of liver enzymes. Liver damage can be detected biochemically by the increase in serum levels of sorbitol dehydrogenase (SDH), serum glutamic oxaloacetate transaminase (SGOT) and serum glutamic pyruvic transaminase (SGPT) activities (enzymes characteristic for the liver).

In one study,[20] researchers investigated the *in vivo* effect of oral acetyl-boswellic acids on galactosamine/endotoxin induced liver damage in mice. The data indicate that pretreatment with the boswellic acids one hour prior to injection with the bacterial endotoxin, prevented liver injury as evidenced by the activities of liver enzymes. In the control, (untreated animals), galactosamine/endotoxin induced liver injury, the SGPT, SGOT and SDH levels were increased dramatically, by almost 10 times the normal levels, within eight hours after noxious challenge. Pretreatment with boswellic acids significantly reduced the increase in the levels of these biochemical indices as shown in the following (Figs. 3.23, 3.24, 3.25).

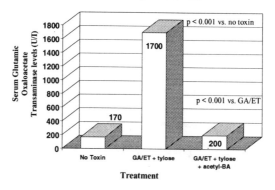

Fig. 3.23

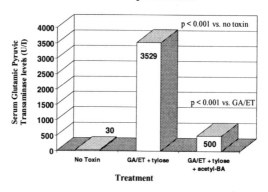

Fig. 3.24

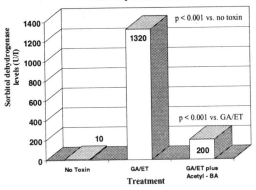

Protection by oral acetyl-Boswellic Acids (acetyl-BA)
pretreatment against Galactosamine/Endotoxin (GA/ET)
induced hepatitis in mice

(Ref. 20)

Fig. 3.25

IV. Antihyperlipidemic and anti-atherosclerotic activity

Boswellia serrata as well as *Commiphora mukul* or *wighti* belong to the same family of plants, Burseraceae. In fact, gum resin is a product from both plants which can be utilized medicinally. The fraction of *Commiphora mukul,* Gugulipid,® a well-known lipid-lowering agent, is closely related to the Salai guggal gum of *Boswellia serrata.* The anti-hyperlipidemic and anti-atherosclerotic activities of Salai guggal gum have been evaluated.

Atal et al [27,28,104] tested the antihyperlipidemic activity of the oleogum resin in animals fed a high-fat diet. The whole oleogum resin of *Boswellia serrata* in a dose of 500 mg/kg and its alcoholic extract in a dose range of 25-50 mg/kg was administered orally to the experimental animals. This regimen resulted in 30-50% decrease in serum cholesterol levels and 20-60% decrease in triglyceride levels.

As illustrated in (Fig. 3.26), the alcoholic extract of Salai guggal administered orally to different animals in a dose of 100 mg/kg decreased cholesterol/triglycerides levels by 42/60%, 45/48% and 25/62%, in rats, in weaned rats and in cockerels, respectively.

46

Effect of Salai guggal treatment (100 mg/kg defatted ethanolic extract
(SGEE)), on serum cholesterol and serum triglyceride levels
in hyperlipidemic animals

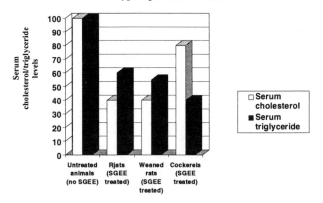

(Ref. 14)

Fig. 3.26

In another study to monitor the anti-atherosclerotic activity of extracts of Salai guggal,[87] rabbits were fed with a diet containing cholesterol and saturated fat for three months. They were divided into four groups. Treatment was started on day 50 and on day 90, in the first two groups, respectively, and continued up to day 150. The other two groups served as untreated controls. Serum cholesterol and triglycerides were monitored at different intervals and were found to decrease during the treatment. The results indicate the possible role of Salai guggal extracts in the prevention and reversal of atherosclerotic processes.

The mechanism for the cholesterol lowering effect of Salai guggal extracts has been elucidated using an experimental model of cholesterol synthesis in rats. The incorporation of a radio-labelled acetate into a cholesterol molecule was used to evaluate the possible effect of Salai guggal on the biosynthesis of cholesterol. The presence of labeled acetate in the cholesterol molecule decreased markedly in both *in vitro* and *in vivo* studies in the presence of Salai guggal,[29,30,100] revealing the cholesterol lowering effect.

V. Anti-ulcerogenic activity

One of the most common side effects encountered with the use of anti-inflammatory drugs is stomach upset, resulting from gastric irritation, which if left unattended may lead to ulcerative disease. Aspirin is probably the best known example of an anti-inflammatory drug, recognized for both its unquestionable efficacy as well as frequently reported side effects in the form of gastric irritation.

Any potential anti-inflammatory compound should be tested for its effects on the gastrointestinal tract. Boswellic acids have been shown to inhibit the development of ulcers in fasted rats.[36] The protective effect was dose dependent in a single dose administration study, as shown in (Fig. 3.27).

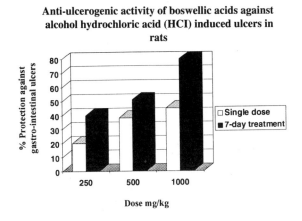

Anti-ulcerogenic activity of boswellic acids against alcohol hydrochloric acid (HCI) induced ulcers in rats

Fig. 3.27

The comparative anti-ulcerogenic activity of boswellic acids against alcohol induced ulcers in rats was evaluated. In a single dose administration of 500 mg/kg of body weight, the mixture of β-boswellic acids was found to exert the maximum protective effect as shown in (Fig. 3.28).[36]

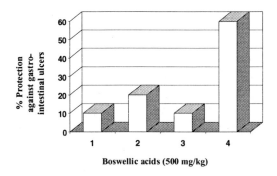

Comparative anti-ulcerogenic activity of boswellic acids against alcohol hydrochloric acid induced ulcers in rats

(Ref. 36)

Fig. 3.28

Boswellic acids used in this study
1. Acetyl hydroxy β-boswellic acid
2. Acetyl 11-keto β-boswellic acid
3. Acetyl hydroxy 11-keto-β-boswellic acid
4. Boswellic acids mixture

2. Clinical studies with Boswellin® and boswellic acids

The most prominent application of Salai guggal is in the management of arthritis. In general, any form of arthritis manifests with more or less evident inflammation, thus an anti-inflammatory drug like boswellic acids has therapeutic potential. It should be noted, however, that different forms of arthritis may respond with a different degree of success to any anti-inflammatory treatment. Before any treatment is implemented, a thorough medical diagnosis of the condition should be obtained.

Arthritis is a painful and disabling condition expressed in a variety of clinical forms. The most common forms of arthritis are osteoarthritis, gout and rheumatoid arthritis.

Osteoarthritis causes a degeneration of the joint cushion material, cartilage. There is a link between being overweight and bio-mechanical stress on joint surfaces, resulting in the degeneration of the joints. Osteoarthritis is the most common joint disease, affecting some 40 million Americans, 85% of whom are above the age of 70 years. Losing weight and a moderate exercise program can prevent and slow down the progress of this disease.

Gout is a genetic metabolic disease resulting from the excess production of uric acid. It precipitates in joints and tendons in the crystal form of monosodium urate, causing acute inflammation and pain of the joint and most often the joint of the big toe. A majority of the sufferers are men. Gout is aggravated by overindulgence in food, alcohol, chronic lead poisoning, emotional stress and infections.

Rheumatoid arthritis affects about 1% of the population mostly young adults, and predominantly women. The origin of this disease is unknown, but the immune system of the body turns against its own connective tissue, causing debilitating inflammation of the joints and internal organs. Boswellic acids were found to be effective in the treatment of arthritis in several separate clinical trials.

I. Early human study

The earliest human studies with boswellic acids in the treatment of arthritis were conducted in 1981-82, based on the positive results heretofore obtained in animal studies.

Pachnanda *et al*[14,107] conducted clinical trials on 175 volunteer patients suffering from rheumatoid arthritis and ankylosing spondylitis of a moderate to severe type.

The population studied consisted of a group of males and females, 5 to 75 years old, suffering from arthritis for a period of 1 to 5 years, with little or no relief from conventional anti-arthritic medications. The most predominant clinical symptoms reported by the patients included: morning stiffness of joints, pain, loss of grip strength and difficulty in perfoming routine tasks. Some of the patients were bedridden due to their poor physical condition.

All patients received *Boswellia serrata* extract in a dose of 200 mg three times a day, continuously for at least 4 weeks. Of the 175 patients, 122 (70%) experienced relief from the morning stiffness within 2-4 weeks of the initiation of treatment. Of the 122 patients who had clinical improvement, 17 were switched to a placebo treatment and showed recurrence of symptoms associated with arthritis within 10 days.

Of the remaining 53 cases, 35 showed some improvement and the other 18 (10%) had no appreciable improvement within a week after starting the treatment. None of the patients complained of any side effects in the cases of treatment with *Boswellia serrata* extract.

The results of the clinical evaluation of boswellic acids in the study are shown in the pie graph (Fig. 3.29) as a percentage of patients with various responses to the treatment. Results were graded as excellent, good, fair and poor:

Clinical effectiveness of boswellic acids in the treatment of arthritis

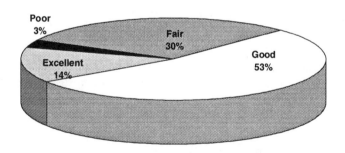

(Ref. 14)

Fig. 3.29

II. Open field clinical evaluation

An open trial on the anti-arthritic properties of boswellic acids was performed at the Medical College, Patiala, India.[11,98] The study was carried out over eight weeks on 30 open field cases (patients received only boswellic acids without a control group). The diagnosis and classification of the disease was based on the American College of Rheumatology criteria.

Clinical conditions were assessed based on the duration of morning stiffness, fatigue time measured as the time elapsed between waking up and when the patient felt tired, time required to walk 25 feet, articular index or degree of joint tenderness, joint swelling and grip strength in both hands. The clinical evaluation was calculated as the mean arthritic score from the above-listed symptoms. In addition, the red blood cell sedimentation rate (ESR) was evaluated in the study group. The ESR establishes the rate at which red blood cells settle down (sediment) in the test tube. The higher the ESR value, the more active the arthritic condition.

The results of this study are shown in (Fig. 3.30 and 3.31).

Open field study on 30 Subjects with rheumatoid arthritis

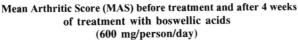

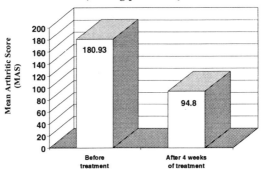

Fig. 3.30

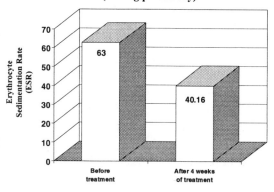

Mean value of Erythrocyte Sedimentation Rate (ESR) before
treatment and after 4 weeks of treatment with boswellic acids
(600 mg/person/day)

Fig. 3.31

III. Double blind clinical evaluation

A double blind trial employed 30 cross-over cases (patients
received boswellic acids and a matching placebo; after a period of
time the treatments were switched) of patients suffering from rheumatoid arthritis. Boswellic acids were administered orally in a dose
of 200 mg three times a day, and the control group received a matching placebo containing lactose. The mean arthritic score in the group
receiving boswellic acids came down after four weeks from the
pretreatment value of 238.4 ± 17.35 to 94.67 ± 24.72; the mean
ESR value was reduced from 65.93 ± 7.7 to 49.2 ± 7.4. As a result
of substituting boswellic acids with a placebo (cross-over), the
arthritic score rose again after 4 weeks of the regimen from $125.6
\pm 22.72$ to 181.06 ± 24.2; the ESR increased from 38.06 ± 7.9 to
45.13 ± 8.6. The results are shown in (Figs. 3.32, 3.33, 3.34, 3.35).

Double blind study on 30 subjects with rheumatoid arthritis

Mean Arthritic Score (MAS) before treatment and after 4 weeks of treatment with boswellic acids (600 mg/person/day)

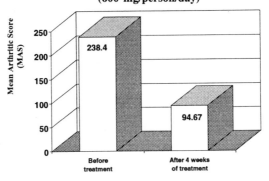

Fig. 3.32

Mean value of Erythrocyte Sedimentation Rate (ESR) before treatment and after 4 weeks of treatment with boswellic acids (600 mg/person/day)

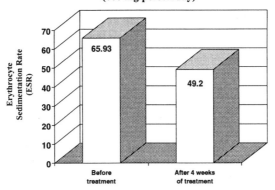

Fig. 3.33

**Mean Arthritic Score (MAS) before treatment and 4 weeks
after cross-over to placebo**

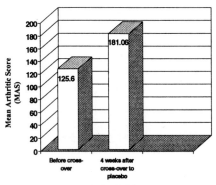

Fig. 3.34

**Mean value of Erythrocyte Sedimentation Rate (ESR) before
treatment and 4 weeks after cross-over to placebo**

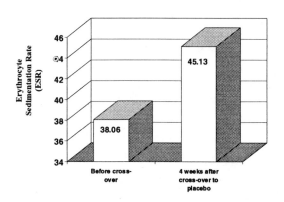

(Ref. 11)

Fig. 3.35

None of the patients in the open field or double blind study showed any side effects except for two patients, who reported a minor skin reaction, which disappeared on discontinuation of the treatment. For the first 10-14 days, the patients needed an additional analgesic to alleviate the active disease-associated pain. These results indicate that boswellic acids should be considered as potential anti-arthritic therapeutic agents which can improve the course of the disease within a few weeks.

IV. Open field clinical evaluation of a topical analgesic

In an open field study, a topical cream Chilisin,® consisting of boswellic acids, capsaicin and methyl salicylate was evaluated in 12 volunteers monitored by several independently working chiropractors. The study was coordinated by Synergy Wellness Clinics, DeSoto, Texas. The preliminary results of the ongoing study are hereby reported:

The criteria for admission to the study group included a long standing arthritic condition manifested by one or all of the following symptoms: chronic joint pain, diminished joint mobility, joint swelling, heat sensation over the involved joint and morning stiffness. All willing and qualified participants were asked to sign the informed consent, previously approved by the chiropractor supervising the study. Each patient was provided with a two weeks supply of Chilisin® cream, and asked to apply the cream daily as needed. In addition, each patient agreed to fill out the self-evaluation questionnaire before the treatment started and again after two weeks of treatment.

Based on the information obtained from 12 patients in the study, seven women and five men, middle-aged to elderly, Chilisin® cream applied regularly on a daily basis for at least two weeks resulted in significant improvement in all parameters tested, with pain score being diminished by half as a result of the treatment (Fig. 3.36). Of the twelve participants, five evaluated the therapeutic effects of Chilisin® as very good, three as good, three as modest and one as not satisfactory. The person who claimed a "not satisfactory" result of the treatment was an 83-year-old woman who preferred "Blue Ice" over Chilisin.® None of the participants reported any side effects of the treatment. Two patients complained of a specific odor of the cream (due to the presence of methyl salicylate).

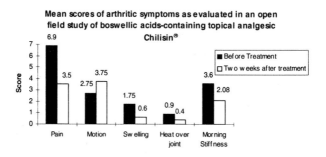

Fig. 3.36

V. Clinical uses in veterinary medicine

While arthritis is a disabling condition in humans, for animals like horses, arthritis has been an indication for a probable death sentence. In one such case, described by Dr. McDonagh, doctor of osteopathy, a racing horse was literally saved from its death sentence due to treatment with boswellic acids.[23] This horse had a stifle problem due to arthritis, which was keeping it from competitive racing.

Several veterinarians tried numerous treatments without success, and as a last-ditch effort, the horse was treated with boswellic acids. According to Dr. McDonough this horse recovered after 30 days of treatment and returned to competitive racing.

Another case was described by Patrick T. Maloney, D.V.M. of Kentucky, who treated a horse for chronic post-operative arthritis of the knee with boswellic acids.[32] According to Dr. Maloney, the painful condition was alleviated after just a few days of the treatment. He also noted that this specific condition was refractory to the previously attempted conventional therapy with non-steroidal anti-inflammatory drugs.

Ronald Blackwell, D.V.M. of Kentucky, and Gary Kaufman, D.V.M. of Arizona, reported the usefulness of boswellic acids in chronic inflammatory conditions in horses, such as stifle problems, sore backs, bowed tendons and bone spurs.[32]

These reports of use of boswellic acids in veterinary practice resulted in its introduction to the American Association of Equine Practitioners in Orlando, Florida, and prompted several veterinarians to utilize boswellic acids as an anti-inflammatory compound alternative to steroids.

The extract of *Boswellia serrata* appears to be a potent natural therapeutic agent in the treatment of rheumatoid arthritis. Its novel mode of action, through the selective inhibition of leukotriene formation, and lack of side effects so often associated with frequently recommended NSAIDs, makes it a strong candidate for consideration in the management of inflammatory disorders.

TOXICOLOGY AND SAFETY ASPECTS

The oleogum resins of *Boswellia,* under the common name "olibanum", are approved by the FDA for food use (21 CFR 172.510) and were included by the Council of Europe (1974) in the list of substances, spices and seasonings deemed "admissible for use, with a possible limitation of the active principle in the final product."[84,85]

Acute and chronic toxicity studies

The toxicity study of *Boswellia* gum resin established the acute dermal and oral LD_{50} at more than 5 gm/kg of body weight in rats and rabbits.[84] The oral and intraperitoneal LD_{50} was greater than 2 gm/kg in rats and mice.[87]

The four week toxicity study done on rats and primates, as well as a three month toxicity study done on rabbits, showed that oral administration of boswellic acids at 5 to 10 times the ED_{50} value did not result in side effects.[88]

In a separate four week toxicity study performed on rats, the animals fed with *Boswellia* gum resin at doses of 0.5 and 1.0 gm/kg body weight showed no significant change in blood biochemistry and morphology, including the histopathology of major organs.[6]

A study on female rats receiving boswellic acids orally, in a dose ranging from approximately 50 to 250 times the calculated human therapeutic dose, for a period of six months, revealed no significant changes in biochemical and hematological parameters as compared to untreated controls[112] (Fig. 4.1).

Effect of treatment with Boswellic Acids (for a period of six months) on biochemical parameters in rats

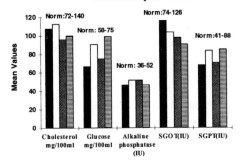

Effect of treatment with Boswellic Acids (for a period of six months) on biochemical parameters in rats

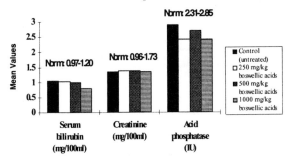

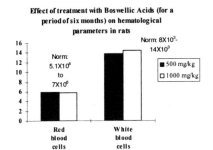

(Ref. 112)

Fig 4.1

60

Ulcerogenic index of *Boswellia serrata* extract

An important consideration of any potential anti-inflammatory agent should be its effect on the gastrointestinal mucosa and its possible role in ulcer formation.

A comparison of the ulcerogenic index of boswellic acids and the commonly used anti-inflammatory drug phenylbutazone, was performed. Results of this comparative study on mice are presented in (Fig. 4.2).[5]

Ulcerogenic Index of Boswellic Acids (BA) and Phenylbutazone (PB)

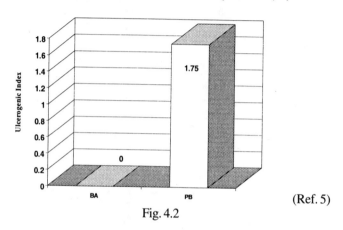

(Ref. 5)

Fig. 4.2

In a separate study, rats were fed with the oleogum resin for six weeks at doses of 0.5 and 1.0 gm/kg of body weight. This regimen did not result in any untoward ulcerogenic effect as confirmed by examination of the rat stomachs.[87]

Based on cited studies performed with rodents it may be concluded that unlike phenylbutazone, boswellic acids do not induce gastric ulcerations.

Toxicology of boswellic acids topical application

An important application of *Boswellia* extracts is in products for topical use such as soaps, detergents, creams, lotions and perfumes. The concentrations commonly used in these products are as follows:[84]

	Soap	*Detergent*	*Cosmetic creams, lotions*	*Perfume*
Usual	0.03%	0.003%	0.005%	0.12%
Maximum	0.2%	0.02%	0.03%	0.8%

In the evaluation of topical toxicity, olibanum gum was applied at the maximum topical concentration to intact or abraded rabbit skin for 24 hours. The preparation was found to be moderately irritating.[88] When tested at a concentration level of 8% in petrolatum, it produced no irritation after an eight hour closed-patch test on human subjects.[88] Cases, however, of mild allergic reactions to incense obtained from *Boswellia* species have been reported.[89] The concentration of boswellic acids in Chilisin® a topical analgesic and anti-inflammatory cream, is 5%.

To test for sensitization reactions in dermal applications, a test was carried out on 25 human volunteers.[91,92] The olibanum gum, again tested at the maximum concentration level of 8% in petrolatum produced no sensitization reactions.[90]

Comparison of *Boswellia serrata* extract with known anti-inflammatory drugs

An essential characteristic of any long term treatment is its degree of safety, as measured by lack of side effects. This is of particular importance with regards to boswellic acids, since most of the inflammatory conditions, like arthritis, require long term treatment. The following table lists the most common over-the-counter anti-inflammatory and anti-arthritic drugs[111] and compares their safety with that of the boswellic acid preparation found in Boswellin.®

Evaluated drug	GI Ulceration	Other GI effects	Skin	Hematology	Liver	Kidney	CNS	Other effects
IBUPROFEN	++*	+++	+	-	-	-	+	+ (pruritus)
INDOMETHACIN	+++	++++	+	+	-	+	++	+(respiratory)
KETOPROFEN	++	+++	+	-	-	-	-	None
NAPROXEN	+	++	+	-	-	-	++	None
MEFENAMIC ACID	+	++	+++	-	-	-	-	None
PHENYLBUTAZONE	+	++++	-	+++	+	+	+	+ (Cardiovascular)
ASPIRIN	+++	++++	+	+	+	+	+	+ (Respiratory)
BOSWELLIC ACIDS[6,16,112]	-	-	-	-	-	-	-	None

*The incidence of individual side effects is graded according to approximate percentage, namely,

+	0.1-5.0%
++	5.0-10%
+++	10-15%
++++	> 20%

From the results of the toxicological studies described in this review, it is evident that boswellic acids are safe for oral as well as topical use.

ANALYTICAL PROFILE

Although a total of more than one hundred compounds have been reported in the various species of *Boswellia*, the two triterpenoid acids, O-acetyl-β-boswellic acid and β-boswellic acid, are considered the most characteristic.[24]

The isolation and structure elucidation of the pharmacologically active triterpenoid acids obtained from the gum resin of *Boswellia serrata* has been performed using sophisticated techniques such as mass spectrometry and the study of NMR spectra. These studies revealed the presence of triterpene acids, as described in the section on the chemistry of *Boswellia serrata*.

Samples of *Boswellia* extracts vary widely in terms of types and levels of the triterpene acids present. These Δ^{12} α-amyrin acids were first isolated and characterized by Winterstein and Stein.[59] The earliest work on boswellic acids from *Boswellia serrata* was performed by Pardhy and Bhattacharya,[64,65] who used chromatographic methods to purify the extracts and mass spectral data to characterize the β-boswellic acids.

A non-aqueous titrimetric method was developed by Gupta et al[71] for the estimation of total triterpene acids in *Boswellia serrata* extract. The total triterpene acids estimated by this method include β-boswellic acid, acetyl-β-boswellic acid, 11-keto-β-boswellic acid and acetyl-11-keto-β-boswellic acid.

The analysis was performed on the basis of β-boswellic acid, which constitutes more than 30% of the total triterpene acids. Estimation of triterpene acids alone or in combinations of two was done using functional group analysis. The functional groups analyzed were acetyl and hydroxy groups at the 3-position and the keto group at the 11-position.

The analytical profile of a sample of *Boswellia serrata* extract, determined on the basis of the methods elucidated by earlier researchers, is outlined in the following pages:

1. COMPOSITION

Boswellia serrata extract, as Boswellin,® contains about 65% boswellic acids, which are pentacyclic triterpenoid acids. The major constituents of boswellic acids are:

i)   β-boswellic acid
ii)   11-keto-β-boswellic acid

It also contains small amounts of α-boswellic acid and γ-boswellic acid.

2. DESCRIPTION

2.1 Nomenclature

Chemical names of the major constituents of *Boswellia serrata* extract

    i) β-boswellic acid: 3-Hydroxyurs-12-en-23-oic acid

    ii)11-keto-β-boswellic acid: 3-Hydroxyurs-11-one-12-en-23-oic acid

2.2 Formulae of the constituents of *Boswellia serrata* extract

2.2.1 Empirical formulae

    i) β-boswellic acid $C_{30}H_{48}O_3$

    ii)11-keto-β-boswellic acid $C_{30}H_{46}O_4$

2.2 Structural Formulae

    i) β-boswellic acid

    ii)11-keto-β-boswellic acid

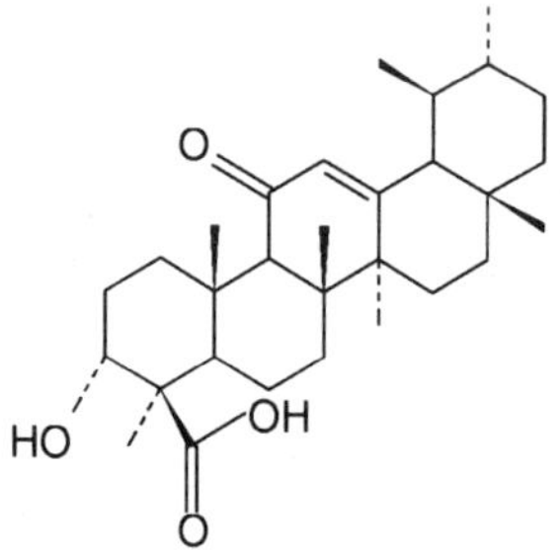

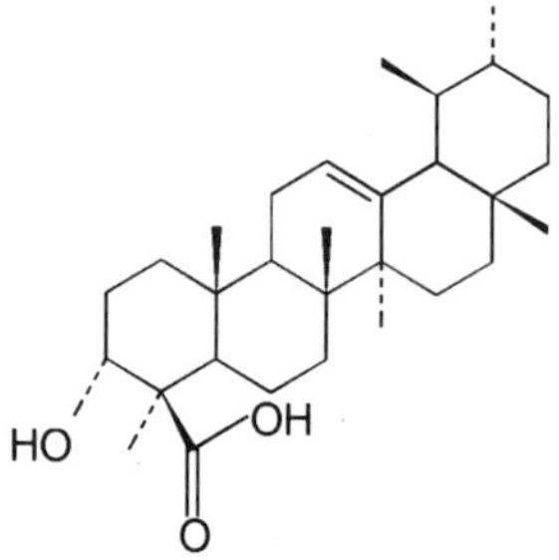

2.3 Molecular weights of the major constituents of *Boswellia serrata* extract

    i) β-boswellic acid   456.71

    ii)11-keto-β-boswellic acid   470.69

2.4 Appearance, color and odor

Pale yellow to light brown amorphous powder with a characteristic odor.

3. PHYSICAL PROPERTIES

3.1 Solubility

Soluble in alcohol, chloroform and in acetone. Insoluble in water.

3.2 Spectra

The infrared spectrum of *Boswellia serrata* extract obtained as potassium bromide dispersion, is given in (Fig. 5.1). The spectrum was recorded by a Shimadzu Fourier Transform Infrared (FTIR) 8101A Spectrophotometer. The spectral assignments in correlation with β-boswellic acid are given below:

a) -OH stretching vibrations 3435 cm^{-1}

b) -CH$_3$, -CH$_2$ stretching vibrations 2928 cm^{-1}
2872 cm^{-1}

c) C = 0 (of -COOH) stretching vibrations 1703 cm^{-1}

d) -C = C (unsaturation) stretching vibrations 1662 cm_{-1}

e) -CH$_3$, -CH, bending vibrations 1456 cm^{-1}

INFRARED SPECTRUM OF BOSWELLIA SERRATA EXTRACT

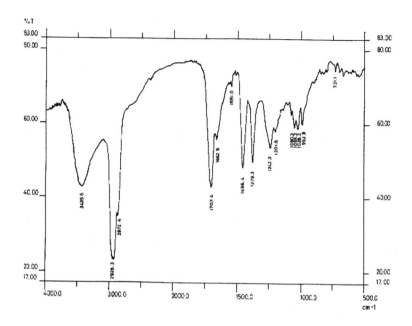

Fig. 5.1

3.2.2   Ultraviolet Spectrum

The ultraviolet spectrum of *Boswellia serrata* extract in methanol in the region of 200nm to 350nm exhibits one maximum at 248nm and one minimum at 228nm. The ultraviolet spectrum is given in (Fig. 5.2)

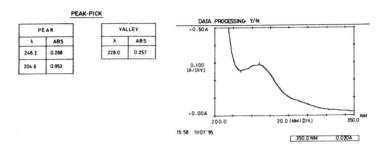

Fig. 5.2

3.3 Melting Range

Boswellia serrata extract melts between 100°C and 110°C.

3.4 Specific Rotation

Boswellia serrata extract exhibits specific rotation of about +65° in chloroform.

3.5 Moisture Content

Boswellia serrata extract shows moisture of about 3%. This is only surface moisture and is determined by the Karl Fischer method.

3.6 pH

pH of 1% w/v suspension of *Boswellia serrata* extract is about 4.5.

⊙

4. METHODS OF ANALYSIS

4.1   Chromatographic Analysis

4.1.1   Thin Layer Chromatography (TLC)
The following TLC procedure is specific only for identification of boswellic acids.

Mobile Phase

Methylene chloride: Acetone (98:2)

Stationary Phase
Precoated silica gel plate of GF 254

Sample Preparation
Dissolve about 50mg of the sample in 10ml of methanol

Working Standard Preparation
Dissolve about 50mg of standard in 10ml of methanol

Procedure
Spot separately 10μl of sample preparation and working standard preparation. Develop the chromatography to 3/4 of the plate. Remove and dry the plate in a current of air.
Detection
Spray the plate with antimony (v) chloride spray reagent* and dry the plate at 120°C till the spots appear.
Limit: The profile of the sample preparation is the same as that of the working standard preparation.

Preparation of antimony (v) chloride: Mix one volume of antimony (v) chloride with four volumes of chloroform.

4.1.2 High Performance Liquid Chromatography (HPLC)

Boswellia serrata extract can be quantified by the following HPLC method:

In this method, a *Boswellia serrata* extract with 70% boswellic acids is used as the standard. (For determination of boswellic acids, only the major peaks obtained in the sample are compared with that of the standard.) HPLC Chromatograms of the standard and sample of *Boswellia serrata* extract are given in (Fig. 5.3).

Mobile Phase
A) Mix water and acetonitrile (20:80), filter and degas
B) Acetonitrile Gradient Programming:

Initial	100%	A
20 minutes	100%	B

Standard Preparation
Weigh accurately about 25mg of the working standard *Boswellia serrata* extract and transfer into a 25ml volumetric flask. Dissolve in methanol, dilute to volume and mix.

Sample Preparation
Weigh accurately about 25mg of *Boswellia serrata* extract and transfer into a 25ml volumetric flask. Dissolve in methanol, dilute to volume and mix.

Chromatographic System

The liquid chromatograph is equipped with 250nm UV detector and a 150 x 4.6mm stainless steel column that contains ODS packing (Shimpack from Shimadzu is suitable). The flow rate is adjusted to one ml per minute.

Procedure
Separately inject, in duplicate, equal quantity (20µl) of standard preparation and sample preparation into the chromatograph and measure the responses for the major peaks. Compare the profile of the sample with that of the standard and calculate the content of boswellic acids as follows:

$$\frac{\text{Total area of major peaks in sample x standard weight}}{\text{Total area of major peaks in standard x sample weight}} \text{ X } 70$$

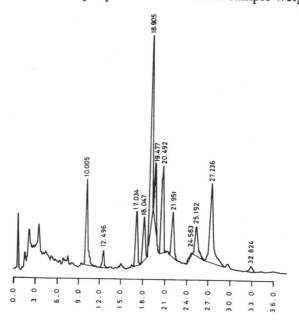

HPLC Chromatogram of *Boswellia serrata* extract (Standard)

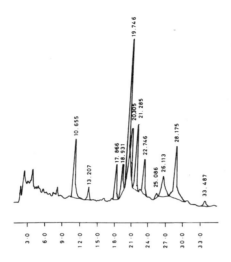

<u>HPLC Chromatogram of *Boswellia serrata* extract</u> (Sample)

Fig. 5.3

4.2   Titration

4.2.1   Acid Base Titration

Boswellic acids in *Boswellia serrata* extract can be estimated by the following method:

Weigh accurately about 0.75g of the sample and transfer to a 250ml conical flask. Dissolve in 75ml neutralized alcohol. Titrate against 0.1N sodium hydroxide using phenolphthalein solution as an indicator. Each ml of 0.lN sodium hydroxide is equivalent to 0.0456g of boswellic acids.

4.2.2   Non-aqueous Titration

Triterpenes are estimated by non-aqueous titration and boswellic acids are triterpenes. Therefore, this method was followed to estimate boswellic acids content.

Weigh accurately about 0.5g of *Boswellia serrata* extract and transfer to a dry conical flask. Dissolve in 50ml of dimethyl formamide and titrate against 0.1N potassium methoxide using thymol blue solution as an indicator. Each ml of 0.lN potassium methoxide is equivalent to 0.0456g of boswellic acids.

REFERENCES

1. *Hager's Handbuch der Pharmazeut. Praxis* (1972) Vol. III, 4ᵗʰ Ed. Springer-Verlag, Berlin, Heidelberg, New York, pp 491.
2. *The Wealth of India: Raw Materials* (1948) Vol I , CSIR Publications, Delhi, pp 208-210.
3. Kirtikar, K.R. and Basu, B.D., *Indian Medicinal Plants*, Vol. I (1935), 521-529.
4. Chopra, R.N, Nayar, S.L., Chopra, I.C. (1956) *Glossary of Indian Medicinal Plants*, CSIR, Delhi.
5. Singh, G.B. and Atal, C.K. (1986) Pharmacology of an extract of salai guggal ex-*Boswellia serrata*, a new non-steroidal anti-inflammatory agent, *Agents and Actions*, 18 (3/4), 407-411.
6. Singh, G.B. *et al.* (1993) Boswellic Acids *Drugs of the Future*, 18(4) 307-309.
7. Reddy, G.K. *et al.* (1987). Effect of a new non-steroidal anti-inflammatory agent on lysosomal stability in adjuvant-induced arthritis. *Italian J. Biochem.* 36, 205-217.
8. Menon, M.K. and Kar. A. (1971) Analgesic and psycho-pharmacological effects of the gum resin of *Boswellia serrata*. *Planta Med.* 19, 338-341.
9. Kakrani, H.K.(1981), Guggul - a review. *Indian Drugs*, 18, 417.
10. *RRL Newsletter*, *Jammu, India*, (1982), 9(3)
11. *Annual Report, Regional Research Laboratory, Jammu, India* (1987-88), pp 1-2.
12. Reddy, G.K. *et al.* (1989) Studies on the metabolism of glycosaminoglycans under the influence of new herbal inflammatory agents. *Biochemical Pharmacology*, Vol. 38(20), 3527-3534.
13. Neilsen, K (1986)*Incense in ancient Israel.* E.J. Brill, Leiden,Chapter 1.
14. Gupta, V.N, Yadav, D.S., Jain, M.P., Atal, C.K. (1987) Chemistry and Pharmacology of the gum resin of B.serrata. *Ind. Drugs* 24(5), 221-231.
15. Safayhi, H., *et al.* (1992). Boswellic Acids: Novel, specific, non-redox inhibitors of 5-lipoxygenase.*J. Pharm. and Exp. Therap.*, 261(3), 1143.
16. *Annual Report, Regional Research Laboratory, Jammu, India*, (1991-92), pp 10-11 and proprietary process by Sami Chemicals & Extracts (P) Ltd., Bangalore, India.
17. Ammon, H.P.T. *et al.* (1993). Mechanism of anti-inflammatory actions of curcumin and boswellic acids, *J. Ethnopharmacology*, 38, 113-119.

18. Dennis, T. J. *et al.* (1980). *Bull. Med. Ethnobot. Res.*, 1, 353.

19. Weiner, M. (1991) On Herbs, *Health Foods Business*, pp 16-17.

20. Safayhi, H. *et al.* (1991) Protection by Boswellic Acids against galactosamine/ endotoxin-induced hepatitis in mice. *Biochem. Pharmacol.* 41(10) 1536-7.

21. Sharma, M.L. *et al.* (1988) Effect of Salai Guggal, ex-*boswellia serrata* on cellular and humoral immune response and leucocyte migration. *Agents and Actions* 24, 161-164.

22. Wagner, H. (1989) Search for new plant constituents with potential antiphlogistic and antiallergic activity. *Planta Med.*, 55, 235-241.

23. McDonagh, E.W. (May 1992), *Acres, U.S.A.*

24. *Patent No. 0 552 657 A1* (1993), European Patent Office.

25. Kumar, A. And Saxena, V.K. (1979). TLC and GLC studies of the essential oil from *Boswellia serrata* leaves, *Indian Drugs*, 16, 80-83.

26. Roonwal, M.L. *et al.* (1960). *Ind. For. Records, Entomol.* 9, 215-39.

27. Atal, C.K. et al. (1981). Salai guggal: a promsing anti-arthritic and anti-hyperlipidaemic agent. *Brit. J. Pharm.* 74, 203

28. Singh, G. B. *et al.* (1984)., *Ind. J. Pharmacol.*, 16, 115.

29. Zutshi, U. *et al.* (1980). *Ind J. Pharm.* 12, 59.

30. Zutshi, U. *et al.* (1980). *13 IPS* (Abstract), Raipur.

31. Zutshi, U. *et al.* (1984) *36 IPC* (Abstract), Bangalore.

32. McCrea, B. (1993). Ancient Indian medical herb proving itself a winner for modern-day equine athletes. *J. Am. Hol. Vet. Med. Assoc.*, 12(1), 19

33. Menon, M.K. and Kar, A. (1969). Analgesic effects of the gum resin of *Boswellia serrata*. *Life Sciences* 8(1), 1023-28.

34. Reddy, G.K., Dhar, S.C., Singh, G.B. (1987) Urinary excretion of connective tissue metabolites under the influence of a new non-steroidal anti-inflammatory agent in adjuvant induced arthritis. *Agents and Actions*, 22, 99-105.

35. Gogate, V.M. *Dravyagunavignyana*, Continental Prakashan, Pune, India

36. Research Report, No. 1001, (1995), *Sami Chemicals and Extracts*, India.

37. Atal, C. K., Symp. Int., Workshop on Pharmacological and Biochemical Approaches on Medicinal Plants, School of Biol. Sciences, Madurai, India, Abstract, cited in (40).

38. Mishra, V. *et al.*, *Bull. Bot. Soc. Univ. Saugar.*, 27, 59, (1980).

39. Mukherji, S. *et al.* (1970). Studies on plant anti-tumor agents. *Ind. J. Pharm.*, 32, 48.

40. *Selected Medicinal Plants of India* (1992), CHEMEXIL, India, pp 65-66.

41. *Sushruta Samhita* (1949) Shree Gulabkunvarba, Vol 1-6, Ayurvedic Soc., Jamnagar.

42. ***Charaka Samhita,*** (1968) Chowkhamba, (2nd ed), Sanskrit Series Office, Varanasi.

43. Nadkarni, A.K. (1976). ***Indian Materia Medica,*** Vol 1, p 211, Popular Prakashan Pvt. Ltd., Bombay, India.

44. Kapoor, L.D. (1990) ***Handbook of Ayurvedic Medicinal Plants,*** CRC Press Inc., 83.

45. Dhar, M *et al.* (1968). Screening of Indian plants for biological activity, Part I. ***Indian J. Exp. Biol.*** 6, 232.

46. Hairfield, E.M. and Hairfield, H.H. (1989) GC, GC/MS and TLC of β-boswellic acid and O-acetyl-β-boswellic acid from *B. Serrata, B. Carterii* and *B. Papyrifera.* ***J. Chromatographic Science*** 27, 127.

47. ***Merck Index*** (1976) An Encyclopedia of Chemicals and Drugs. (9thed.) No. 6679. Merck and Co. Inc. Rahway, NJ.

48. Goswami, M. And Sen, N. (1942),.***Ind. Soap J.*** 27, 173-83.

49. Fowler, C. J and Malandkar, M.A. (1925) ***J. Ind. Inst. Sci.,*** B, 221-39.

50. Siddiqui, M.M.H. *et al.* (1984). Studies on Kundur, the oleogum resin of *Boswellia serrata* ***Nagarjun,*** 28(1&2), p 1-3.

51. Siddiqui, M.H. *et al.* (1984). Chemical standardization of Kundur, 50.

52. Pearson, R.S and Singh, P. (1918). ***Ind. Forest Records,*** 6, 321.

53. Simonson, J.L. and Owen, L.N. (1949) ***The Terpenes,*** 2.

54. Roberts, O.O, (1923). ***J. Soc. Chem. Ind. (Lond.),*** 42, 486.

55. Guenther, E.S., (1943).***Am. Perfumer,*** 45, 41-43.

56. Girgune, J.B. and Gar, B.D. (1979) ***J. Sci. Res. (Bhopal, India),*** 1, 119.

57. Ferdinado, T., (1937). ***An. Chim. Appli.,*** 27, 178-88.

58. *Ruzicka, L. et al. (1944).* **Helv. Chim. Acta,** 27, 1859-67.

59. Winterstein, A. and Stein, G. (1932), Untersuchungen in der Saponinreihe. Zur Kenntnis der Mono-oxy-triterpensauren. Hoppe-S Z. ***Physiol. Chem.,*** 208, 9-25.

60. Saha, A.N. and Hiyogi, R.K. (1958), Surfactants from resin acids (Part I) ***Ind. Soap J.,*** 23, 167-171.

61. Shamma, M. *et al.* (1962). ***J. Org. Chem.*** 27, 4512-17.

62. Budzikiewiz, H. *et al.* (1963).***J. Am. Chem. Soc.*** 85, 3688-99.

63. Pardhy, R. S and Bhattacharya, S.C. (1978) Structure of serratol, a new diterpene cembranoid alcohol from *Boswellia serrata* Roxb. ***Ind. J. Chem. Sec. B.,*** 16B, 171-173

64. Pardhy, R. S and Bhattacharya, S.C. (1978) Tetracyclic triterpene acids from the resin of *Boswellia serrata* Roxb ***Ind. J. Chem. Sec. B.,*** 16B, 174-175.

65. Pardhy, R. S and Bhattacharya, S.C. (1978). Pentacyclic triterpene acids from the resin of *Boswellia serrata* Roxb *Ind. J. Chem. Sec. B.*, 16B, 176-178.

66. Beton, J.L. *et al.* (1956). *J. Chem Soc.*, 2904.

67. Gupta, V.N. *et al.* (1987). Condensation reactions of the methyl ester of 3-keto-beta-boswellic acid with aromatic aldehydes. *Indian Drugs*, 25(2), 70-72.

68. El Khadem, H. *et al.* (1993). Derivatives of Boswellic Acids, *J. Ethnopharmacology*, 38(2-3), 113-9.

69. Safayhi, H. *et al.* (1994). Structure requirements for 5-LO inhibition by Boswellic Acids, *European J. Pharm. Sci.*, v-2(1-2), 101.

70. Product Literature, *Sabinsa Corporation, U.S.A.*

71. Gupta, V.N. *et al.* (1984). *Indian Drugs*, 21,526.

72. *Bull. Imp. Inst. Lond.* (1919), 17, 159.

73. Malandkar, M.A. (1925), *J. Ind. Inst. Sci.*, 3, 240-243.

74. Sharma, R.A. and Varma, K.C. (1980). Studies on gum obtained from *Boswellia serrata*. *Ind. Drugs*, 17, 225

75. Ammon, H.P. 1, Mack, T., Singh,G.B., Safayhi, H. (1991) Inhibition of leukotrienes B₄ formation in rat peritoneal neutrophils by an ethanolic extract of the gum resin exudate of *Boswellia serrata*. *Planta Med.* 57 (3), 203-7.

76. Bhuchar, V. M *et al.*, (1982). Constituents of gum exudate obtained from *Boswellia serrata*. *Ind. J. Tech.*, 20(1), 38

77. Bhargava, G. G., (1978). Studies on the chemical composition of salai gum, *Ind. For.*, 104, 174

78. Simpson, J. C. E. and Williams, N. E. J. *J. Chem. Soc.* 686 (1938).

79. Simpson, J. C. E. and A. R. K. George *J. Chem. Soc.* 793 (1941).

80. Garg, S. C., *Ind. J. Pharm.*, 36 -46, (1974).

81. Sharma, M.L. , Bani, S., Singh, G.B. Anti-arthritic activity of boswellic acids in bovine serum albumin (BSA)-induced arthritis. *Int. J Immunopharmacol* 1989,11(6), 647-52.

82. Reddy, G.K., Dhar, S.C., Singh, G.B. Biochemical investigations of a new non-steroidal anti-inflammatory agent in adjuvant induced arthritis in relation to serum glycohydrolases and glycoproteins. *Leather Sci* 1986, 33, 192-9.

83. *Annual report*, (1987-88), *Regional Research Laboratory, Jammu,India*, p 42.

84. Opdyke, D.L.J. (1978), Fragrance raw materials monograph, olibanum gum. *Fd. Cosmet.Toxicol.* 16, 837.

85. Council of Europe (1974). Natural Flavoring Substances, Their Sources, and Added Artificial Flavoring Substances. Partial Agreement in the Social and Public Health Field. List N(1), Ser. 1(b), no. 93, p 48, Strasbourg.

86. Atal, C.K. *et al.* (1982) *XV IPS (Abstract)*, Chandigarh.

87. Singh, G.B. *et al.* (1984) Symposium "Recent Advances in Mediators of Inflammation and anti-inflammatory agents," Abstract, *Regional Research Laboratory, Jammu, India*, p. 38.

88. Weir, R.J. (1971). *Report to RIFM,* 25 August, cited in Ref. 84.

89. Basto, A.S., Azenha, A. (1991). Contact dermatitis due to incense. *Contact Dermatitis,* 24(4), 312-313.

90. Kligman, A.M. (1971). *Report to RIFM,* Sept. 27 cited in Ref. 84.

91. Kligman, A.M. (1966). The identification of contact allergens by human asssay. III. The maximization test-procedure for screening and rating contact sensitizers. *J. Invest. Derm.,* 47, 393.

92. Kligman, A.M. and Epstein,W. (1975). Updating the maximization test for identifying contact allergens, *Contact Dermatitis,* 1,231.

93. Arctander, S. (1960). *Perfume and flavor materials of natural origin.* Arctander, Elizabeth, NJ, pp 463-64.

94. Duwiejua, M. *et al.* (1993). Anti-inflammatory activity of resins from some species of the plant family Burseraceae. *Planta Med.* 59, pp 12-16.

95. Kar, A. (1977). Effect of the gum resin of *Boswellia serrata* on the cardiovascular system and isolated tissues. *Ind. Drugs and Pharm. Ind.,* July-Aug., 1-4.

96. Guenther, E. (1950). *The Essential Oils,* Vol 4, pp 352, Van Nostrand Inc. Princeton, N.J.

97. Kapil, A. And Moza, N. (1992). Anti-complementary activity of Boswellic Acids - an inhibitor of C-3 convertase of the classical complement pathway. *Int. J. Immunopharmac.,* 14 (7), pp 1139-1143.

98. Singh, G.B. *et al.* (1992). New phytotherapeutic agent for treatment of arthritis and allied disorders with novel mode of action. *IV and Int. Congress on Phytotherapy, Sept. Munich, Germany, Abstract SL 74.*

99. Afaq, S.H. and Siddiqui, M.M.H. (1984). Pharmacology and Clinical Studies on Unani Medicinal Plants Vol 1. Kundur (*Boswellia serrata*) and Guggal (*Commiphora mukul*). A.K. Tibbiya College, Aligarh Muslim University, Aligarh, India.

100. Zutshi, U. *et al.* (1986). Mechanism of cholesterol-lowering effect of salai guggal. *Ind. J. Pharm.* 18(3), pp 182-3.

101. Kulkarni, R.R. *et al.* (1991). Treatment of osteoarthritis with a herbomineral formulation , a double blind placebo controlled study. *J. Ethnopharmacology,* 33(1-2), 91-95.

102. Kulkarni, R.R. *et al.* (1992). Efficacy of an Ayurvedic formulation in rheumatoid arthritis, a double blind placebo controlled study. *J. Ethnopharmacology*, 34(2), 98-101.

103. Yadav, D.S. *et al.* (1986). Synthesis of β-boswellic acid analogues. *36th IPC, Bangalore, India*, A 12.

104. Atal, C.K. *et al.* (1980).Salai guggal ex-*Boswellia serrata* a promising antihyperlipidemic and antiarthritic agent. *Ind. J. Pharm.*, **12,59.**

105. Singh, G.B. *et al.* (1984) Symposium "Recent Advances in Mediators of Inflammation and anti-inflammatory agents," Abstract, *Regional Research Laboratory, Jammu, India*, p 49.

106. Singh, G.B. *et al.* (1984) Symposium "Recent Advances in Mediators of Inflammation and anti-inflammatory agents," Abstract, *Regional Research Laboratory, Jammu, India*, p 40.

107. Pachnanda, V.K. *et al.* (1981). *Ind. J. Pharm.*, 13, 63.

108. Mathew, K.M. (1983). *The flora of Tamil Nadu and Carnatic*, The Rapinat Herbarium, St. Joseph's College, Tiruchirapalli, India, Part I, p. 223.

109. (Schechter, P.J. *et al.* (1987) in *Inhibition of Polyamine Metabolism* (McCann, P.P., Pegg, A.E. and Sjoerdsma, A. eds.) pp. 345-364. Academic Press.

110. Huang, M.T. *et al.* (1988) *Cancer Res.* 48; 5941

111. Rainsford, K. (1984) Side effects of anti-inflammatory analgesic drugs: epidemiology and gastrointestinal tract. *TIPS,* 156-159.

112. Research report No. 78 (1995). *Sami Chemicals and Extracts* (India).